Student's Guide to Masterton and Slowinski's
CHEMICAL PRINCIPLES *Third Edition*

USING THE INTERNATIONAL SYSTEM OF UNITS

RAYMOND BOYINGTON
Department of Chemistry,
University of Connecticut

WILLIAM L. MASTERTON
Department of Chemistry,
University of Connecticut

SAUNDERS GOLDEN SUNBURST SERIES

1977   W. B. SAUNDERS COMPANY / Philadelphia / London / Toronto

W. B. Saunders Company: West Washington Square
Philadelphia, PA 19105

1 St. Anne's Road
Eastbourne, East Sussex BN21 3UN, England

1 Goldthorne Avenue
Toronto, Ontario M8Z 5T9, Canada

The front cover illustrations, *Variation within a Sphere, No. 10*, a metal sculpture by Richard Lippold, courtesy of The Metropolitan Museum of Art.

Student's Guide to Masterton and Slowinski's CHEMICAL PRINCIPLES USING THE INTERNATIONAL SYSTEM OF UNITS   ISBN 0-7216-1903-7

© 1977 by W. B. Saunders Company. Copyright 1973 and 1975 by W. B. Saunders Company. Copyright under the International Copyright Union. All rights reserved. This book is protected by copyright. No part of it may be reproduced, stored in a retrieval system, or transmitted in any form or by any means, electronic, mechanical, photocopying, recording, or otherwise, without written permission from the publisher. Made in the United States of America. Press of W. B. Saunders Company.

Last digit is the print number: 9 8 7 6 5 4 3 2 1

PREFACE

This study guide has been written as an aid to the student of general chemistry — to supplement a textbook or lecture series, or to guide independent study. The chapter sequence and selection of topics parallel that of *Chemical Principles, Using the International System of Units*, 4th edition, by W. L. Masterton and E. J. Slowinski.

Each chapter begins with a set of questions. These are meant to suggest some of the things you should be looking for in studying the topics of the chapter and the corresponding text and lecture material. You should add your own questions to this partial listing. Next follows a list of concepts and mathematical manipulations with which you should already be familiar in order to get the most out of your study. A chapter summary then presents an overview of some of the ideas treated in this unit of study.

In addition, each chapter of the guide contains a section entitled Basic Skills. Here are listed the concepts and problem-solving techniques that must be mastered before you can work problems in the text or on an examination. Each skill is illustrated by an example, either directly or by reference to an appropriate example and corresponding problems in the text. Illustrative problems in the text are set off by key words which indicate the skill applied.

Next is the Self-Test, which consists of true-false and multiple choice questions (each with one best answer) as well as problems which generally apply more than one of the basic skills. There are also one or two "bonus" problems, set off by asterisks (*), which we hope you will find intriguing. Excluding these special problems, the entire Self-Test should require between 60 and 90 minutes to work. Each test is similar to the examinations we have given at Storrs over the last six years. In fact, many of the questions and nearly all the problems are taken from these exams.

Finally, following answers to the Self-Test, is a list of recommended readings. These may be used to supply more basic, background study, interesting and important applications, and in-depth study as well.

The major objective in writing this guide has been to provide you with tests by which you can measure your understanding; to help you apply what you learn to new problems and suggest new avenues of study; and to help you acquire a feel for the chemist's view of the physical world. We welcome your suggestions for the many ways there must be for making this a more useful guide.

SUGGESTIONS FOR THE USE OF THIS GUIDE

First, let's agree that what you get out of this study of general chemistry depends in great part on what you are willing to put into it. What follow are a few suggestions as to how you can make the most of your effort.

There are certain basic skills that you will frequently need to rely on, possibly further develop. The most fundamental of these are reading with comprehension and solving mathematical problems once they have been set up. There is probably no better way to learn or improve a language skill than through repeated usage, and that is one reason for this guide. Also, the problems and the tests in this guide will help you to decide just what you do and do not understand. As for the math, there are likely to be very few calculations in this course which you cannot do. These may consist of working with exponential notation or logarithms; solving first order (linear) equations; or solving less frequently encountered second order (quadratic) equations. For these you may simply need some review. Suggestions for background readings follow; additional help is generally yours for the asking, from fellow students and from your instructor.

1. Attend lectures and take notes. Notes are important in answering questions such as these: What seem to be the important points? What is the lecturer's emphasis? What new examples and applications are given? What kind of experiments are done? *Think* about what is being said, do not merely take notes.

2. Prepare for the lecture! Read over the material you expect to be discussed. Try to get a general idea of what the material is about. (Some lecturers assume a basic preparation, some do not. Even if the lecture should tell you little new, you will have learned more by having done the work yourself.) The more you know about what the lecturer is saying, the more you will be able to learn; the fewer notes you will need to take, and the better they will be; the more questions you can raise; or, the more time you can devote to other studies.

3. Get as much as you can out of discussions with other students and with your instructor, out of discussion or problem-solving classes — by being prepared, and by understanding (first by discovering for yourself) the inherent limitations and potentials of your class. (You will generally find, for example, that an instructor is better able to answer specific questions than he is to guess what problems you are having.) Make your problems known, as

early and as clearly as possible. Take advantage of the skills and insights of your instructor and your fellow students! You usually do have a very real influence on how beneficial a discussion class may be. Remember that this guide is designed to help you evaluate your progress for yourself, to recognize those areas where you need help, and to assist you in finding it; but it is no substitute for the classroom.

4. Review regularly and frequently. This is where your notes should be most helpful, and hopefully this guide as well. Keep this use in mind when taking notes from lecture or text. Look for connections between topics.

5. Read more chemistry — whether out of curiosity or from a feeling of helplessness. The more you do, the more rewarding this study must be.

To sum up:

Before attending lecture: Skim through the textbook chapter, reading introductory paragraphs and section headings, looking at tables and graphs; read and *add your own notes* to the first three sections of the study guide chapter. Then go back and *work* through the textbook chapter. In this and all other parts of the course, you've got to *actively participate.* Try to *anticipate* answers; write in questions you need answered; jot down your own conclusions in your own words.

Before discussing problems: Work through as many problems as possible in this guide and in your text; work the self-test in the guide; note anything requiring further explanation; think about your answers — do they seem reasonable in terms of what you already know? Do they suggest other problems?

Before taking exams: Review the main concepts and their physical meaning by referring to your notes and to the chapter summaries; recall and try to anticipate the emphases. Work some more problems.

For further assistance or enjoyment: Ask questions of your fellow students and your instructor. Do some reading. Teach what you have learned.

ADDITIONAL STUDY AIDS AND SUGGESTED READINGS

Math Preparation and Problem-Solving Manuals

Butler, I. S., *Relevant Problems for Chemical Principles*, Menlo Park, Calif., W. A. Benjamin, 1970.

Gibson, G. W., *Mastering Chemistry: A Problem Solving Guide for Introductory Chemistry*, Philadelphia, W. B. Saunders, 1975.

Masterton, W. L., and Slowinski, E. J., *Elementary Mathematical Preparation for General Chemistry*, Philadelphia, W. B. Saunders, 1974.

O'Connor, R., *Solving Problems in Chemistry*, New York, Harper & Row, 1974.

Peters, E. I., *Problem Solving for Chemistry*, 2nd ed., Philadelphia, W. B. Saunders, 1976.

Pierce, C., *General Chemistry Workbook*, San Francisco, W. H. Freeman, 1971.

Risen, W. M., Jr., *Problems for General and Environmental Chemistry*, New York, Appleton-Century-Crofts, 1972.

Sienko, M. J., *Chemistry Problems*, Menlo Park, Calif., W. A. Benjamin, 1972.

Programmed Instruction and Instruction in Programming

Barrow, G. M., *Understanding Chemistry*, New York, W. A. Benjamin, 1969.
Soltzberg, L., *BASIC and Chemistry*, Boston, Houghton Mifflin, 1975.

Journals and Magazines

Chemical and Engineering News
Chemistry
Journal of Chemical Education
Scientific American

Audio-Visual Aids

Tapes, films, slides, programmed instruction materials and other resources that might be used for self-instruction are often available but not used. Find out where they are and how to use them, and whether they are worth your time. Often available at any library. Reprints of published papers are often available free from the author; sometimes, for a small fee, from the publisher.

Consider:

O'Connor, R., *Topics-Aids: A Guide to Instructional Resources for General Chemistry*, Washington, D. C., American Chemical Society, 1975.

General References and Popular Accounts

Encyclopedia of Chemical Technology, New York, Wiley, 1970.

Rochow, E., *Modern Descriptive Chemistry*, Philadelphia, W. B. Saunders, 1977.

What's Happening in Chemistry?, Washington, D. C., American Chemical Society, 1976 (an annual publication).

Woodburn, J. H., *Taking Things Apart & Putting Things Together*, Washington, D. C., American Chemical Society, 1976.

Handbooks and Compilations of Data

Handbook of Chemistry and Physics, Cleveland, Chemical Rubber.
Lange's Handbook of Chemistry, New York, McGraw-Hill Book Co.

Foreign-Language Texts

Consider the translations of the text and lab manual of Masterton and Slowinski:

Chimie: Théorique et Expérimentale, G. S. Gantcheff (translator), Montreal, Les Editions HRW Ltée, 1974.
Experimentelle Einführung in Grundlagen und Methoden der Chemie, D. Krug (translator), Stuttgart, Gustav Fischer Verlag, 1976.

The authors wish to thank Elaine Rossi for typing the manuscript.

CONTENTS

1
CHEMISTRY: AN EXPERIMENTAL SCIENCE 1

2
ATOMS, MOLECULES, AND IONS 13

3
CHEMICAL FORMULAS AND EQUATIONS 25

4
THERMOCHEMISTRY 43

5
THE PHYSICAL BEHAVIOR OF GASES 59

6
THE ELECTRONIC STRUCTURE OF ATOMS 75

7
THE PERIODIC TABLE AND THE PROPERTIES OF ELEMENTS ... 89

8
CHEMICAL BONDING 103

9
PHYSICAL PROPERTIES AS RELATED TO STRUCTURE 121

10
AN INTRODUCTION TO ORGANIC CHEMISTRY 133

11
LIQUIDS AND SOLIDS; PHASE CHANGES 145

12
SOLUTIONS 161

13
WATER, PURE AND OTHERWISE 173

14
SPONTANEITY OF REACTION; ΔG AND ΔS 185

15
CHEMICAL EQUILIBRIUM IN GASEOUS SYSTEMS 199

16
RATES OF REACTION 217

17
THE ATMOSPHERE 229

18
PRECIPITATION REACTIONS 239

19
ACIDS AND BASES 253

20

ACID-BASE REACTIONS 267

21

COMPLEX IONS; COORDINATION COMPOUNDS 283

22

OXIDATION AND REDUCTION; ELECTROCHEMICAL CELLS 299

23

OXIDATION-REDUCTION REACTIONS; SPONTANEITY AND
EXTENT 311

24

NUCLEAR REACTIONS 325

25

POLYMERS, NATURAL AND SYNTHETIC 337

1
CHEMISTRY: AN EXPERIMENTAL SCIENCE

QUESTIONS TO GUIDE YOUR STUDY

1. What role does chemistry play in today's world? How might this role change in the near future? (Can you recall any chemical issues recently debated in the news media?)
2. What kinds of problems do chemists try to solve? Are there any specially useful or simplifying approaches to their solution? Is there a chemist's "point of view"?
3. What kinds of materials does the chemist generally work with in the laboratory? (Are they, for example, the materials of the "real" world, such as steel, plastic, beer, and tobacco smoke?)
4. How does the chemist obtain, prepare, and purify the materials he studies?
5. What are some of the instruments and techniques of the chemist?
6. What properties or changes in properties does the chemist measure and use to describe materials? How are these measurements communicated to other scientists and to the public?
7. What limitations and uncertainties are there in these measurements? How are these communicated?
8. What kind of test would you perform in your kitchen to show that a sample of baking soda is "pure"? (What is the chemical meaning of "pure"?)
9. What are some of the unsolved problems — the frontiers — in chemistry?

10. What questions would you, a reporter for a local newspaper, want to ask of a government-employed chemist?

11.

12.

YOU WILL NEED TO KNOW*

Concepts

Though no previous encounter with chemistry is assumed at this point, at least a general understanding is assumed for the notions of matter, energy, composition, experiment, measurement . . . .

Math

1. How to use exponential notation — Appendix 4.
2. How to recognize and solve first order (linear) equations, such as $y = ax + b$ — see the Basic Skills section as well as the Readings.

CHAPTER SUMMARY

You know, perhaps all too well, that we live in an age of science and technology, an age that is now two to three centuries old yet very much with us. For example, nearly ninety per cent of all scientists that there have ever been are alive now. At least through the first half of this century science seemed, by any measure, to be growing exponentially: about every ten to fifteen years the number of scientific journals approximately doubled and the number of compounds known to chemists likewise doubled. (Try estimating the number of papers in chemistry expected in the year 2000, knowing that there were about 50 000 in 1950.) How long can this rapid growth continue? Are there signs that it has already begun to taper off?

This course of study will introduce you to the underlying principles and methods of one of the more fruitful areas of scientific endeavor, an endeavor involving many thousands of persons in many nations.

But just what does a chemist do? Most of us would probably agree that a chemist is one who is qualified to determine the feasibility of chemical

---

*All chapters and appendices referred to are in *Chemical Principles*; otherwise, see the readings list for appropriate background material.

change. A major objective of the chemist is to be able to predict the conditions under which a chemical reaction may occur and to describe the course the reacting system may take. (Does this suggest the kind of social role the chemist might play? After more than ten years of research, and applying a well known chemical principle, F. Haber was able to describe the conditions under which atmospheric nitrogen could be converted to ammonia. A process made commercial in 1913, the Haber synthesis now accounts for most of the ammonia produced, whether for eventual use in fertilizers, explosives, or Everybody's Ammonia Cleanser.)

The history of modern chemistry really began with the introduction of quantitative experimentation, with measurement. The idea of composition, barely mentioned here, is taken up in more detail in the next two chapters. There are several levels of meaning to *composition*: one speaks of atomic composition, for example; or elemental composition; or composition by mass and so forth. All of these usages refer to the way in which component parts or building blocks, whether simple in themselves or very complex, are put together to form matter as we know it.

A major objective for the student in an introductory course is to see how the principles of chemistry are experimentally established and then applied in predictions of reaction feasibility. Consequently, the chemical systems that we will look at will be simple in composition and simple in the number of things that happen.

BASIC SKILLS

In this introductory chapter you are expected to become familiar with some of the common types of measurements that chemists make and the experiments they carry out to separate and identify substances. You should, for example, be familiar with the measurement of mass and the units in which masses of substances are commonly expressed. Again, you should understand the principles that govern such separation techniques as distillation and chromatography. In another area, you should be able to explain how such properties as boiling point and absorption spectrum can be used to identify a substance and check its purity.

A few specific skills are required in this chapter. They include the following.

1. Apply the rules of significant figures to calculations based upon experimental measurements.

These rules are illustrated both in the body of the text and in Examples 1.2 and 1.3. Students ordinarily have little trouble learning these rules, but they often forget to apply them in carrying out chemical calculations. None of the problems at the end of Chapter 1 refer directly to significant figures,

4 • 1–Chemistry: An Experimental Science

but in each case you are expected to report an answer to the correct number of significant figures. This will be true throughout the text and this study guide.

2. Use the conversion factor approach to convert lengths, volumes, masses, or other measured quantities from one unit to another.

This is a general approach which can be applied to a wide variety of problems in chemistry. It will be used extensively in later chapters of the text. If you are not familiar with it, this chapter gives you an excellent opportunity to become skilled in its use.

The principle behind the conversion factor approach is illustrated in Examples 1.4 and 1.5 in the text. Notice (Example 1.4) that it is readily applied to conversions involving more than one step. The example below further illustrates this application.

A certain American car is reported to have a fuel economy of 18.0 miles/gal (U.S.). Convert this to kilometres per litre.

Clearly, two conversions are required; miles must be converted to kilometres and gallons to litres. The first conversion factor is available directly from Table 1.1:
  1. 1 mile = 1.609 km

The other factor does not appear directly in the table. However, the conversion may be accomplished by first converting gallons to quarts:
  2. 1 gal = 4 qt

and then to litres:
  3. 1ℓ = 1.057 qt (U.S.)

With these conversion factors available, we can set up the problem:

$$18.0 \,\frac{\text{miles}}{\text{gal}} \times \underset{(1)}{\frac{1.609 \text{ km}}{1 \text{ mile}}} \times \underset{(2)}{\frac{1 \text{ gal}}{4 \text{ qt}}} \times \underset{(3)}{\frac{1.057 \text{ qt}}{1 \ell}} = 7.65 \,\frac{\text{km}}{\ell}$$

Note that each conversion factor is set up in such a way (1.609 km/1 mile) that the original unit (miles) cancels out, leaving the desired unit (km).

Many of the problems at the end of Chapter 1 require conversions of one type or another. In Problem 1.2, for example, you are asked to carry out a conversion which is exactly the reverse of the one just worked. Problems 1.11, 1.14, 1.25, and 1.28 illustrate other straightforward conversions from one unit to another. Problem 1.32 is an interesting example of how the conversion factor approach can be applied, at a somewhat more advanced level, to a practical problem of considerable environmental importance.

Basic Skills • 5

3. Use an algebraic equation to solve for an unknown quantity, given or having calculated all the other quantities in the equation.

This is another basic skill which you will use frequently in general chemistry. In this chapter it is illustrated in simple form by
— *temperature conversion (°C, °F, K) carried out using Equations 1.2 and 1.3.* Note Example 1.1 and Problems 1.3, 1.12a, and 1.26a at the end of the chapter.
— *relating density, mass, and volume, using the defining equation:*

$$\text{density} = \frac{\text{mass}}{\text{volume}} \; ; \; D = \frac{m}{V}$$

This equation is used in Example 1.3 and in the following example.

What is the volume of a sample of aluminum (density = 2.70 g/cm^3) weighing 12.0 g?

Here, we rearrange the defining equation to solve for volume. To do this, we first multiply both sides of the equation by V to obtain

$$D \times V = m$$

and then divide both sides by the density, D:

$$V = \frac{m}{D}$$

Substituting the quantities given in the statement of the problem:

$$V = \frac{12.0 \text{ g}}{2.70 \text{ g/cm}^3} = 4.44 \text{ cm}^3$$

Several of the problems at the end of Chapter 1 (1.9, 1.12, 1.22, 1.23, 1.26) require that you be familiar with the density relation.

4. Use a table of solubilities (Table 1.8) to design an experiment to separate two solids by fractional crystallization.

The general approach followed here is described in the text discussion and applied in Problems 1.17 and 1.31. In working these problems note that
— solubilities at a given temperature can be treated like conversion factors. Thus, to work 1.17a, we might consider that, at 80°C, 98 g tartaric acid ≏ 100 g water.

— the solubility of one solid is assumed to be independent of the presence of the other. Thus, at 100°C, it is assumed that 100 g of water will dissolve 98 g of tartaric acid, regardless of how much succinic acid is present, and 71 g of succinic acid, regardless of how much tartaric acid is present. Obviously, this is an approximation, but it is probably more nearly accurate than any other simple assumption we might make at this point.

SELF-TEST

True or False

1. Changes in physical state, like melting and boiling, tend to ( ) resolve matter into pure component substances.

2. Physical properties may be used to characterize, identify, a ( ) substance.

3. In trying to identify a certain liquid compound "L," a ( ) student finds that its density, freezing and boiling points, absorption spectrum, and behavior in a chromatography column are indistinguishable from those of known compound "Z." The student can safely assume that L and Z are one and the same compound.

4. Nearly all countries have already adopted or are now ( ) adopting a metric system as the single recognized system of measurement.

5. The simplification in using the metric system is that, when ( ) converting units within the system, a decimal point is moved.

6. Appropriate conversion factors would allow you to convert ( ) from a volume measurement in cubic feet to a density measurement in grams per cubic centimetre.

7. Since silicon is the second most abundant element in the ( ) earth's crust, it must be one of the cheapest to buy from a chemical supplier.

8. Sodium chloride and potassium dichromate, both solids ( ) which are soluble in water, can probably be separated by taking advantage of differences in solubility.

9. A litre is almost the same as a quart. ( )

10. In dividing 4.053 by 2.46 your calculator displays the ( ) answer 1.647 560 976. You should report 1.647 560 976 as your answer.

Multiple Choice

11. The number 0.002 070 should be written in exponential ( )
notation as
    (a) 2.07×10^{-3} (b) 2.070×10^{-3}
    (c) 2.070×10^{3} (d) 20.7×10^{-4}

12. If radioactive dating reveals the age of a manuscript to be ( )
between 1150 and 1390 a, its age may be represented as
    (a) 1.270×10^{3} a   (b) 1.27×10^{3} a
    (c) 1.3×10^{3} a   (d) 1×10^{3} a

13. How many significant figures are there in the number 6.50×10^{3}? ( )
    (a) two   (b) three
    (c) four   (d) six

14. Consider the four measurements for the mass of a certain ( )
metal coin: 3.0 g, 3.006 g, 3.01 g, and 0.003 002 kg. The average
mass, in grams, should be reported as
    (a) 3.0045   (b) 3.004
    (c) 3.00   (d) 3.0

15. One way to definitely show that gasoline is a mixture of ( )
substances would be to
    (a) measure its density
    (b) measure the temperature during boiling
    (c) burn it
    (d) filter it

16. A procedure appropriate to the separation of the com- ( )
ponents of gasoline is
    (a) burning   (b) fractional crystallization
    (c) fractional distillation   (d) paper chromatography

17. The fact that behavior on melting may allow you to decide ( )
whether or not a sample of matter is a pure substance depends on the
general observation that
    (a) melting points can be measured with high accuracy
    (b) all pure substances can be melted
    (c) the melting temperature is noticeably changed by small amounts of impurities
    (d) the melting point does not vary with sample size

18. A sample of matter that exhibits uniform behavior during ( )
all physical changes would have to be
    (a) an element   (b) a compound
    (c) a mixture   (d) a pure substance

8 • 1–Chemistry: An Experimental Science

19. A suitable, nondestructive test for determining whether or ( )
not a particular beautiful green gem is an emerald would be to
    (a) determine its melting point
    (b) see if it is scratched
    (c) measure its absorption spectrum
    (d) weigh it

20. *Identification* of the solid amino acid components in Super ( )
Crunchy Corn Flakes would probably involve
    (a) distillation   (b) recrystallization
    (c) filtration   (d) chromatography

21. For a given substance, density ordinarily increases in the ( )
order
    (a) solid, liquid, gas   (b) liquid, solid, gas
    (c) gas, liquid, solid   (d) solid, gas, liquid

22. A steel screw weighs 0.50 oz. Its mass in grams (454 g = 1 lb ( )
= 16 oz) is
    (a) 0.50 × 454 × 16   (b) 0.50 × 454/16
    (c) 0.50 × 16/454   (d) none of these

23. If the volume and mass measurements on a sample of ( )
arsenic are 2.10 cm^3 and 12.040 g, the reported value for the density
of arsenic (g/cm^3) should have how many significant figures?
    (a) two   (b) three
    (c) four   (d) five or more

24. How many times larger is the kelvin than a Fahrenheit ( )
degree?
    (a) 1.8   (b) 1/1.8
    (c) 32   (d) they are the same

25. The solid objects A and B are at temperatures of 274 K and ( )
276 K, respectively. When they are placed in contact with each
other,
    (a) heat "flows" from A to B
    (b) heat "flows" from B to A
    (c) the temperature of A drops
    (d) the temperature of B rises

26. The number of kilocalories that is equivalent to 435 ( )
kJ is
    (a) 0.009 62   (b) 104
    (c) 435   (d) 1820

27. The temperature is expected to *change* during the boiling of ( )
a(n) _____ in an open beaker.
    (a) element   (b) compound
    (c) pure substance   (d) mixture

28. Which property should depend on the amount of a pure ( )
substance?
   (a) temperature
   (b) density
   (c) solubility
   (d) volume

29. Which of the following does not involve a chemical change? ( )
   (a) combustion of natural gas
   (b) cooking an egg
   (c) boiling water
   (d) photosynthesis

30. It is found that the chirping frequency (min^{-1}) of a tree ( )
cricket, f, is approximately related to the Celsius temperature by the
equation °C = 0.1f + 8. One would expect that the chirping frequency and the Fahrenheit temperature would be
   (a) unrelated
   (b) equal to each other
   (c) related by the equation for a straight line
   (d) directly proportional to each other

Problems

31. The diameter of a red blood corpuscle is about 8.6 × 10^3 nm. If 2950 corpuscles were laid side by side, how long a straight line would be formed? Give the length in metres.

32. Two substances, A and B, have the following solubilities in g/100 g of water:

	20°C	100°C
A	10	60
B	20	50

   (a) How much water will be required, at 100°C, to dissolve a sample containing 30 g of A and 10 g of B?
   (b) If the solution in (a) is cooled to 20°C, how much A will crystallize out? How much B?

33. Planet X has oceans of liquid ammonia (mp = −78°C, bp = −34°C). The temperature scale used on Planet X takes the melting point of ammonia to be 0°X and the boiling point to be 100°X. That is:

```
-34  —              —  100



-78  —              —   0



  °C                   °X
```

Express 0°C in °X. _____

*34. Suppose you wish to prepare a water solution containing 60 g of A and 20 g of B. The solubility of A, in g/100 g H_2O, depends on the Celsius temperature, t, according to the equation

$$S(A) = 20 + 4.0 \times 10^{-1} t + 2.0 \times 10^{-3} t^2$$

while that of B is given by

$$S(B) = 10 + 3.0 \times 10^{-1} t + 4.0 \times 10^{-3} t^2$$

(a) What minimum amount of water should you use to dissolve the sample at 100°C?
(b) To what temperature can the solution be lowered before B starts to crystallize out?
(c) How much solid A appears at this temperature?

*35. "Die Chromatographie erhielt ihren Namen durch den Umstand, dass die Lagen der bei den Versuchen erhaltenen Zonen ursprünglich durch die Farben der erhaltenen Substanzen identifiziert wurden. Moderne Chromatographie-Methoden werden üblicherweise bei Substanzen angewandt, die farblos sind. Können Sie sich drei Methoden einfallen lassen, die dazu benützt werden können, um die Lagen der Zonen von farblosen Substanzen erkenntlich zu machen?"

SELF-TEST ANSWERS

1. T (The basis, for example, of fractional distillation.)
2. T (These include properties like density, melting point, and solubility.)
3. T (Identity of properties *must* mean identity of composition.)
4. T
5. T

---

*Problems with asterisks are written with the hope of offering a challenge.

Self-Test Answers • 11

6. **F** (Volume and density are not equivalent measurements.)
7. **F** (It is not very readily isolated from its compounds. The distinction between abundance and availability, discussed in Chapter 7, is of great practical importance.)
8. **T** (The basis for fractional crystallization.)
9. **T** (You will find it helpful to *know* at least one conversion factor relating English and metric units of distance, volume, and mass.)
10. **F** (Round off to the correct number of significant figures! Report 1.65.)
11. **b**
12. **c** (The second digit is uncertain by one unit; the third, even more.)
13. **b** (The exponential notation shows that the 6.50 contains these digits.)
14. **d** (The least accurate result is known only to within 0.1 g.)
15. **b** (A constant temperature would be characteristic of a pure substance, as a general rule.)
16. **c** (The bulk of the mixture consists of liquid substances.)
17. **c**
18. **d** (Either a compound or an element would behave this way.)
19. **c**
20. **d** (See Chapter 25.)
21. **c** (Water is exceptional; see Chapters 11 and 13.)
22. **b**
23. **b**
24. **a**
25. **b** (And the temperature of A rises, the temperature of B drops.)
26. **b**
27. **d** (Go back to Questions 1 and 15 above.)
28. **d**
29. **c** (Rather, a physical change.)
30. **c** (Try substituting $0.1f + 8$ for the Celsius temperature in $°F = 1.8°C + 32$.)
31. $2950 \times 8.6 \times 10^3 \text{ nm} \times \dfrac{1 \text{ m}}{10^9 \text{ nm}} = 0.025 \text{ m}$
32. (a) $30 \text{ g A} \times \dfrac{100 \text{ g water}}{60 \text{ g A}} = 50 \text{ g water}$ (more than enough to dissolve 10 g of B)

    (b) A left $= 50 \text{ g water} \times \dfrac{10 \text{ g A}}{100 \text{ g water}} = 5 \text{ g A}$; 25 g A crystallizes

    B left $= 50 \text{ g water} \times \dfrac{20 \text{ g B}}{100 \text{ g water}} = 10 \text{ g B}$; no B crystallizes

33. $°X = a°C + b$; $100 = -34a + b$
    $0 = -78a + b$
    $a = 100/44 = 2.27$
    $b = 177$
    $°X = 2.27°C + 177$; $0° C = 177°X$

12 • 1–Chemistry: An Experimental Science

*34. (a) At 100°C, S(A) = 20 + 40 + 20 = 80; S(B) = 10 + 30 + 40 = 80. To dissolve 60 g of A, use 100 × (60/80) = 75 g water, which is more than enough to dissolve 20 g of B.
  (b) S(B) = 20 × (100/75) = 27 = 10 + 0.30t + 0.0040t^2; 0.0040t^2 + 0.30t − 17 = 0; solving by the quadratic formula, t = 38°C.
  (c) S(A) = 20 + 0.40(38) + 0.0020(38)^2 = 38 g/100 g water. In 75 g water, 38 × 0.75 = 28 g of A remains; 32 g of solid A appears.

*35. The quotation is from Slowinski, E. J., W. L. Masterton, W. C. Wolsey, and D. Krug (translator), *Experimentelle Einführung in Grundlagen und Methoden der Chemie*, Stuttgart, Gustav Fischer Verlag, 1976. Answers might include:
  — Spray with some substance which reacts to give coloration.
  — Check to see if fluorescence occurs with exposure to ultraviolet light.
  — "Label" the mixture components with a radioactive isotope, then map the radioactivity on the chromatogram.
  German is highly recommended to science majors.

SELECTED READINGS

Analytical tools of the chemist are considered in:

Davis, J. C., Introduction to Spectroscopy, *Chemistry* (October 1974), pp. 6–10.
Keller, R. A., Gas Chromatography, *Scientific American* (October 1961), pp. 58–67.
Storms, H. A., Probing Concentration Zero, *Chemistry* (March 1973), pp. 6–10.

Chemistry past, present, and future; careers in chemistry:

Adcock, L. H., Chemistry 200 Years Ago, *Chemistry* (September 1975), pp. 14–15.
Facts and Figures for the Chemical Industry, *Chemical and Engineering News* (June 7, 1976), pp. 33–68.
Handler, P., Whither American Science?, *American Scientist* (July/August 1974), pp. 410–416.
Hill, B. W., Careers in the Chemical Industry, *Chemistry* (November 1975), pp. 6–9.
Impact: Interview with George Pimentel, *Journal of Chemical Education* (April 1974), pp. 224–228.
What's Happening in Chemistry?, Washington, DC., American Chemical Society, 1976.

Problem solving is considered in the readings listed in the preface to this guide; also:

Edman, D. D., Computers and Chemistry, *Chemistry* (January 1972), pp. 6–9.
Masterton, W. L., *Elementary Mathematical Preparation for General Chemistry*, Philadelphia, W. B. Saunders, 1974.

2
ATOMS, MOLECULES, AND IONS

QUESTIONS TO GUIDE YOUR STUDY

1. Can you think of common, everyday observations which suggest that matter is made of atoms and molecules? Or that it isn't?

2. Why are some of these building blocks neutral, while others are charged?

3. How would you experimentally show that sulfur and oxygen combine in a one-to-one mass ratio to form the gaseous compound sulfur dioxide? How would you demonstrate the conservation of matter?

4. Just how small are atoms and molecules? Is there a convenient way of counting them and of expressing their numbers? Of weighing them?

5. If more than one kind of atom exists for a particular element (e.g., strontium-90 is but one kind of strontium atom), then how are we to interpret atomic masses?

6. What holds atoms together in a molecule?

7. What are the supporting arguments, based on observation, for the atomic theory of matter?

8. How do you distinguish, in terms of atomic theory, between an element and a compound? Or, between one iron chloride and another?

9. If the idea of atoms is at least as old as ancient Greece, then why did it take so long for the atomic theorists to get anywhere?

10. If not an atom, then what is the fundamental, ultimate building-block? (Does the question even make sense?)

11.

12.

YOU WILL NEED TO KNOW

See this section of Chapter 1, Study Guide. Also:

Concepts

1. That a chemical symbol may represent one or more atoms of an element; a molecular formula, one or more molecules of a substance — more information is presented in the next chapter.
2. The distinction between element and compound — Chapter 1. (In fact, you are expected to recognize several elements and compounds by name and by formula. But this too will come with repeated use.)

CHAPTER SUMMARY

With more than a little help from his friend the physicist, the chemist has solved the problem of the nature of bulk matter and the changes it undergoes. He interprets ("explains") the unique properties of substances in terms of the unique properties of invisibly small particles that constitute matter: atoms, ions, and molecules. Though invisible, each particle is claimed to possess size and shape, to be endlessly moving, influencing and being influenced by its neighbors. The directly observable behavior of matter is the collective behavior of very large numbers of these particles.* This imaginative view of matter provides the framework for most of the explanation, and success, of modern science. The biologist extends this theory to deal with very large molecules; the geologist, to account for the large-scale process occurring within the earth; the astronomer, to explain the birth and death of stars.

It is a long way to go from an individual molecule of water to a clean mountain brook, or even a beaker of distilled water. Although all the fundamental principles appear to be known, we cannot yet hope to reconstruct the world from first principles. These principles comprise "atomic theory" and "kinetic molecular theory," and were established between 1800 and 1930, approximately. During that time, and since then, the theory has remained dynamic — growing in its detail and sharpness of focus with every new experiment. Atomic theory is the unifying theme of this course.

---

*Relating bulk physical properties to structural units and the forces between them is the subject matter especially of Chapter 9. Explaining chemical properties is what the second half of the text is about.

Now, the chemist certainly doesn't think of atoms as "ultimate" particles; but atoms and, to a certain extent, part of their underlying structure, are as fundamental, as far down the ladder, as he need consider. (The high energy reactions of modern physics, well outside the range of reactions considered to be chemical, have turned up numerous subatomic particles. Except for electrons and nuclei, our only encounter with such interesting and extraordinary matter will be to mention neutrons — and then only to explain the fact that atomic masses must be considered as average values; and to find that low energy, "chemical" neutrons sometimes bring about interesting reactions. More about this latter topic in Chapter 24.)

The arrangements of electrons and nuclei in atoms, ions, and molecules form the theoretical basis of all chemistry. Their properties seem sufficient for explaining the properties of tangible samples of matter. All observable changes are thought of as being changes in the arrangements of electrons and nuclei within atoms, ions, and molecules; as rearrangements of the atoms, ions, and molecules themselves; or as both. (This elaboration of atomic theory begins mainly with Chapter 6 and continues throughout the text.)

The laws of chemical combination (all of the important ones, like the law of constant composition, were known over a century ago) provide some rather indirect support for these ideas. More compelling support for atomic theory, including the existence of particles smaller than atoms, comes from relatively recent experiments: electrical discharge through gases; atomic and molecular spectra (Chapter 6); x-ray diffraction by crystalline solids (Chapter 11); and radioactivity (Chapter 24).

One of the most important things for you to know is the relationship between mass or volume of a sample and the number of particles the sample contains. Volume measurements are typical in the case of handling gases (Chapter 5) or liquids and solutions (Chapter 12); weighing is the only other common measure of an amount of material. (Even the largest molecules are too small to be directly counted: there is no chemical counterpart to everyday counting.) Hence there is the need for relating mass and numbers of particles, the major topic of this chapter.

The gram atomic mass of an element is a convenient reference sample for measuring mass *and* determining the number of particles present in the sample, all at the same time. Several independent methods of determining the number of atoms in exactly 12.000 g of ^{12}C give a value of 6.02×10^{23}. Avogadro's number is to the chemist what the ream of paper is to the typist, and the dozen to the poultry farmer. It is the number of atoms found in one GAM of *any* element; the number of molecules in one gram molecular mass of *any* molecular substance.

Another use for this relationship is seen in the way the chemist finds it convenient to represent reactions; chemical equations, as you will see in the next chapter, show how we imagine counting off reacting particles, and this we will have done by measuring masses.

16 • 2–Atoms, Molecules, and Ions

BASIC SKILLS

1. Use mass analysis data for two or more compounds of two elements to illustrate the Law of Multiple Proportions.

This skill is illustrated in Example 2.1 and Problems 2.1, 2.9, 2.10, 2.25, and 2.26 at the end of Chapter 2. Note that your calculations can be based on a fixed mass (usually one gram) of *either* element.

2. Relate the number of protons and neutrons in an atom to its nuclear symbol.

The nuclear symbol gives the *atomic number* as a subscript at the lower left of the symbol of the element and the *mass number* as a superscript at the upper left.

atomic number = number of protons

mass number = number of protons + number of neutrons

This skill is shown in Example 2.2 and is used to work Problems 2.2, 2.11, and 2.27. In working these problems, note that in a neutral atom the number of electrons is equal to the number of protons; a positive ion is formed by the loss of electrons, a negative ion by a gain of electrons.

3. Given the atomic masses of two elements, calculate the relative (average) masses of their atoms.

What is the ratio of the mass of a strontium atom (AM = 87.62) to that of a beryllium atom (AM = 9.012)? _____
The masses of atoms are in the same ratio as their atomic masses. Hence, a Sr atom must be 87.62/9.012 = 9.723 times as heavy as a Be atom.

See Problem 2.3 at the end of the chapter. The basic concept of atomic mass is also involved, in a somewhat more subtle way, in Problems 2.12 and 2.28.

4. Given the masses and abundances of the isotopes of an element, calculate its atomic mass. Perform the reverse calculation.

The first type of calculation is illustrated in Example 2.3 and used in Problems 2.4, 2.13, and 2.29. The calculation of isotopic abundances from the atomic mass of an element, given the masses of the isotopes, is illustrated in the following example.

Thallium (AM = 204.37) consists of two isotopes, ^{203}Tl (mass = 203.05), and ^{205}Tl (mass = 205.05). What is the % abundance of the heavier isotope?

Let us represent the fraction of the heavy isotope, ^{205}Tl, by x. Since the fractions of the two isotopes must add up to one, that of the light isotope must be $(1 - x)$. The equation for the atomic mass becomes:

AM Tl = 204.37 = x(205.05) + (1 - x)(203.05)

Solving: 204.37 = 205.05x + 203.05 - 203.05x = 2.00x + 203.05

2.00 = 204.37 - 203.05 = 1.32; x = 1.32/2.00 = 0.660

Converting to per cent, we find the abundance of ^{205}Tl to be 66.0%.

Problem 2.14 is entirely analogous to the example just worked. Problem 2.30 is a little more difficult in that three isotopes are involved. Note, however, that you are given the abundance of one of these isotopes.

5. Given either the number of atoms of an element, or the mass in grams of a sample of an element, calculate the other quantity.

This skill is illustrated in Example 2.4, where both types of calculations are involved. Note that the fundamental conversion factor, valid for any element, is

$$1 \text{ GAM} = 6.022 \times 10^{23} \text{ atoms}$$

The numerical value of the gram atomic mass depends upon the nature of the element: 1.008 g for H, 24.30 g for Mg, etc. Thus, for these two elements we have

$$1.008 \text{ g H} = 6.022 \times 10^{23} \text{ H atoms}$$
$$24.30 \text{ g Mg} = 6.022 \times 10^{23} \text{ Mg atoms}$$

Problems 2.5, 2.16, and 2.32 offer simple illustrations of this skill. Problems 2.18 and 2.34 are a little more subtle in that you must also use the

18 • 2–Atoms, Molecules, and Ions

density relationship (D = m/V). Note further that the same skill, along with other conversions, is required in Problems 2.19, 2.21, and 2.37. In Problem 2.19, for example, a logical first step would be to find the number of atoms or gram atomic masses in a kilogram of gold.

6. **Given either the number of molecules of a substance or the mass in grams of a sample of a molecular substance, calculate the other quantity.**

Note the analogy with Skill 5. The basic conversion factor is:

$$1 \text{ GMM} = 6.022 \times 10^{23} \text{ molecules}$$

This factor is used in Example 2.5 to calculate the mass of a water molecule. The reverse calculation is illustrated below.

How many molecules are there in 12.0 g of water, MM = 18.0?_____
To "convert" 12.0 g of water to molecules, we use the conversion factor

$$18.0 \text{ g water} = 6.022 \times 10^{23} \text{ molecules}$$

Hence:

$$\text{no. molecules water} = 12.0 \text{ g} \times \frac{6.022 \times 10^{23} \text{ molecules}}{18.0 \text{ g}} =$$

$$4.01 \times 10^{23} \text{ molecules}$$

This skill is illustrated in its simplest form in Problems 2.6 and 2.17; in both cases, you first have to obtain the molecular mass by adding up the atomic masses of the atoms in the molecule. Problems 2.20 and 2.36 are similar, but involve an additional volume conversion. Problem 2.33 is perhaps the most difficult of this type in that it combines Skills 5 and 6. Note that in using these skills you will need to be familiar with exponential notation (Appendix 4 of the text). Frequently you will find that the numbers you arrive at are either:
— *very large* (number of atoms or molecules in a sample of matter)
— *very small* (mass of an atom or molecule in grams)

SELF-TEST

True or False

1. The chemical properties of an atom are determined by its nuclear charge. ( )

2. In the light of current knowledge, we must view the law of constant composition as only a good approximation to observed behavior. ( )

3. An isotope is one of two or more atomic species having the same atomic number but different numbers of electrons. ( )

4. The following statement is probably consistent with modern atomic theory: a positively charged sodium ion is smaller than a neutral sodium atom. ( )

5. The molecular mass of a substance is simply a number which tells how heavy a molecule is when compared to a chosen reference. ( )

6. The number of molecules in a gram molecular mass of water, 18 g, is 18/12 times the number of atoms in a gram atomic mass of carbon-12, 12 g. ( )

7. All neutral atoms of a given element have the same number of electrons. ( )

8. Stable, bulk samples of matter carry little or no electric charge; that is, they are electrically neutral. ( )

9. The atomic mass of chlorine is 35.5. Chlorine consists of two isotopes of mass numbers 35 and 37. It follows that the Cl-35 isotope must be the more abundant. ( )

10. The atomic mass is an average number that takes into account all known isotopes of an element, including those prepared artificially in the laboratory. ( )

Multiple Choice

11. For the ion, $^{40}_{20}Ca^{2+}$, how many electrons surround the nucleus? ( )
    (a) 18   (b) 20
    (c) 22   (d) some other number

20 • 2–Atoms, Molecules, and Ions

12. Which of the following has the smallest mass? ( )
    (a) an atom of C
    (b) a molecule of CO_2
    (c) one gram of C
    (d) one GAM of C

13. A certain isotope X has an atomic number of 7 and a mass ( ) number of 15. Hence,
    (a) X is an isotope of nitrogen
    (b) X has eight neutrons per atom
    (c) an atom of X has seven electrons
    (d) all of the above

14. The relative abundances of $^{35}_{17}Cl$ and $^{37}_{17}Cl$ in a sample of ( ) elementary chlorine can be determined by
    (a) fractional crystallization
    (b) vapor phase chromatography
    (c) precipitation with silver nitrate
    (d) mass spectrometry

15. To illustrate the Law of Multiple Proportions, we could use ( ) data giving the percentages by mass of
    (a) magnesium in magnesium oxide and in magnesium chloride
    (b) carbon in carbon monoxide and in carbon dioxide
    (c) potassium dichromate in two different mixtures with sodium chloride
    (d) $^{63}_{29}Cu$ and $^{65}_{29}Cu$ isotopes in copper metal

16. Suppose the atomic mass scale had been set up with ( ) calcium, Ca, chosen for a mass of exactly 10 units, rather than about 40 on the present scale. On such a scale, the atomic mass of oxygen would be about
    (a) 64
    (b) 32
    (c) 16
    (d) 4

17. The most direct method for determining atomic masses is ( )
    (a) mass spectrometry
    (b) combining masses
    (c) α-particle scattering
    (d) gas density measurements

18. A sample of an element is ( )
    (a) a collection of atoms with identical numbers of neutrons
    (b) a collection of atoms with identical nuclear masses
    (c) a collection of atoms with identical nuclear charges
    (d) one of the following: air, earth, fire, water

19. Experimental support for the existence of atoms and ( ) subatomic particles includes all of the following except
    (a) electrical discharge through gases
    (b) metal foil scattering of alpha particles

(c) radioactive decay
(d) continuous mechanical subdivision of a single crystal

20. By about how many orders of magnitude (powers of ten) is ( ) the diameter of an atom larger than that of a nucleus?
 (a) 1
 (b) 2
 (c) 4
 (d) 10 000

21. There is always a ratio of small whole numbers between ( )
 (a) the gram atomic and gram molecular masses of an element
 (b) the mass percentages of copper in any two of its compounds
 (c) the masses of copper combined with one gram of element A in CuA and one gram of element B in CuB
 (d) the atomic masses of any two elements

22. The mass of an individual atom is of the order of ( )
 (a) 10^{-22} kg
 (b) 1/2000 g
 (c) 10^{-22} g
 (d) 6×10^{23} g

23. One atom of He weighs 6.63×10^{-24} g. One atom of ( ) oxygen must weigh
 (a) 16.0 g
 (b) 6.02×10^{-23} g
 (c) 2.66×10^{-23} g
 (d) 1.66×10^{-24} g

24. The idea that most of the mass of an atom is concentrated ( ) in a very small core, the nucleus, is a result of the experiments of
 (a) Dalton
 (b) Bohr
 (c) Cannizzaro
 (d) Rutherford

25. The charge on the nucleus of a neon atom is ( )
 (a) +20
 (b) +10
 (c) zero
 (d) −10

26. The reason that the densities of two different gases at the ( ) same temperature and pressure compare as do their respective molecular masses is
 (a) all gases have the same molecular mass
 (b) all gases have the same density at the same temperature and pressure
 (c) all gas molecules are monatomic
 (d) assuming Avogadro to be right, each gas density is directly proportional to the mass of the respective gas molecule

27. Sufficient information for your calculation of the simplest ( ) atom ratio of nitrogen to hydrogen in the compound hydrazine would be

2–Atoms, Molecules, and Ions

(a) the atomic masses of nitrogen and hydrogen
(b) the ratio of masses for nitrogen and hydrogen atoms and the composition by mass of hydrazine
(c) the combining volumes of nitrogen and hydrogen in forming hydrazine
(d) the atomic masses of nitrogen and hydrogen, and the mass composition of several other nitrogen-hydrogen compounds

28. In one gram molecular mass of hydrogen, H_2, there are ( ) Avogadro's number of
   (a) hydrogen atoms   (b) electrons
   (c) hydrogen molecules   (d) neutrons

29. A gram molecular mass of acetic acid, $C_2H_4O_2$, ( )
   (a) contains N molecules   (b) contains 8 × N atoms
   (c) weighs 60 g   (d) all of these

30. The chemical formula, OH^-, represents ( )
   (a) an element   (b) a compound
   (c) a molecule   (d) a diatomic ion

Problems

31. The atomic mass of gallium is 69.72.
   (a) What is the mass in grams of 1.20×10^3 atoms of gallium?
   (b) There are two isotopes of gallium with atomic masses of 68.96 and 70.96. What is the percentage of the lighter isotope?

32. The density of iron is 7.86 g/cm^3. Calculate the volume, in cm^3, occupied by one atom of iron (AM = 55.8; N = 6.02×10^{23}).

33. A molecule of octane contains 8 carbon and 18 hydrogen atoms.
   (a) What is the molecular mass of octane? (AM C = 12.0, H = 1.0)

   (b) How many molecules are there in a sample of octane weighing 26.0 g?

*34. Estimate the density, in g/cm^3, of nuclear matter.

*35. L'élément bore est constitué de deux isotopes de masse 10,02 et 11,01, dont l'abondance est respectivement 18,83 et 81,17%. Calculez le poids atomique du bore.

SELF-TEST ANSWERS

1. T
2. T (We do find that some atom ratios are not ratios of small whole number and that some ratios are at least slightly variable.)
3. F (Different numbers of neutrons.)
4. T (Removing one or more electrons subtracts from the total volume of the atom, which is mostly electrons anyway.)
5. T
6. F (There are N molecules in one GMM, N atoms in one GAM, regardless of the number of grams in either. Rather, the individual structural units have masses that compare 18/12.)
7. T
8. T (An important observation.)
9. T (Can you explain this?)
10. F (Only naturally occurring isotopes.)
11. a (Two electrons have been removed, leaving two units more of positive charge than of negative charge in the atom — 20 protons and 18 electrons.)
12. a (In order of increasing mass: a, b, c, d.)
13. d (c applies to a neutral atom.)
14. d
15. b (The same two elements must be present in the compounds compared.)
16. d (The masses of the individual atoms still compare in the same way: 4/10 is the same as 16/40.)
17. a
18. c (Compare to Question 1.)
19. d (Division could not continue indefinitely. One would ultimately come up against identical units of structure: atoms, molecules, or ions.)
20. c (Atomic, 10^{-8} cm or 0.1 nm; nuclear, 10^{-12} cm.)
21. a (Molecules contain whole numbers of atoms.)
22. c (Divide any GAM by N.)
23. c (An oxygen atom is 16.0/4.00 times heavier.)
24. d
25. b (Same as atomic number.)
26. d (For a given volume, you are comparing equal numbers of molecules.)
27. b (More about atom ratios and chemical formulas in Chapter 3.)
28. c
29. d (If choice b makes no sense, wait till you read Chapter 3.)
30. d

24 • 2–Atoms, Molecules, and Ions

31. (a) $1.20 \times 10^3 \text{ atoms} \times \dfrac{69.72 \text{ g}}{6.02 \times 10^{23} \text{ atoms}} = 1.39 \times 10^{-19}$ g

  (b) $69.72 = 68.96x + 70.96(1 - x); x = 0.620; 62.0\%$

32. $1 \text{ atom Fe} \times \dfrac{55.8 \text{ g}}{6.02 \times 10^{23} \text{ atoms}} \times \dfrac{1 \text{ cm}^3}{7.86 \text{ g}} = 1.18 \times 10^{-23} \text{ cm}^3$

33. (a) $8(12.0) + 18(1.0) = 114$

  (b) $26.0 \text{ g} \times \dfrac{6.02 \times 10^{23} \text{ molecules}}{114 \text{ g}} = 1.37 \times 10^{23}$ molecules

*34. Consider a "typical" AM of 100, and nuclear radius of 10^{-12} cm:
$$D = \dfrac{\text{mass of nucleus}}{\text{vol. of nucleus}} = \dfrac{100 \text{ g}/6 \times 10^{23} \text{ nuclei}}{4/3 \pi (10^{-12})^3 \text{ cm}^3} = 10^{13} \text{ g/cm}^3$$

*35. Even without knowing any French, you may recognize this as Problem 2.4 in the text.

If you are interested, a translation of the text (3d edition) is available. See the Preface to this guide.

SELECTED READINGS

Atomic theory and its history are considered in:

Feinberg, G., Ordinary Matter, *Scientific American* (May 1967), pp. 126–134.
Lagowski, J. J., *The Structure of Atoms*, Boston, Houghton Mifflin, 1964.
Lucretius, *On the Nature of the Universe*, Baltimore, Penguin, 1951.
Patterson, E. C., *John Dalton and the Atomic Theory*, Garden City, N. Y., Doubleday-Anchor, 1970.

Avogadro's number can be determined in several ways; some are considered in:

Hildebrand, J. H., *Principles of Chemistry*, New York, Macmillan, 1964.

Esoteric researches of high-energy physics are considered in Chapter 24 and many recent issues of Scientific American*:*

Glashow, S. L., Quarks with Color and Flavor, *Scientific American* (October 1975), pp. 38–50.

3
CHEMICAL FORMULAS AND EQUATIONS

QUESTIONS TO GUIDE YOUR STUDY

1. What information is conveyed by a chemical formula? A chemical equation? Is there more than one kind of chemical formula or chemical equation?

2. What kind of experiment would you do to show that the formula of water is H_2O, and not HO as Dalton believed? What assumptions, if any, do you need to make?

3. What do you need to know about particular atoms and molecules to be able to write chemical formulas and equations?

4. How does the chemist conveniently deal with large numbers of reacting molecules and with their masses?

5. How do you represent the physical state (solid, liquid, solution . . .) of the materials taking part in a reaction?

6. Are the conditions, such as temperature and pressure, under which a reaction occurs represented by the chemical equation for the reaction?

7. What does a chemical equation tell you about what you would see as a reaction proceeds? (For example, what shape, size, and color crystal is formed in a certain reaction, and how fast?)

8. What does a chemical equation tell you about what the molecules are doing as a reaction proceeds?

9. What happens when you use nonstoichiometric amounts (quantities other than those represented by the chemical equation) in carrying out a reaction?

10. How would you experimentally establish the equation for a reaction?

11.

12.

YOU WILL NEED TO KNOW

Concepts

1. The meaning (and calculation) of gram formula masses (GFM) or, more specifically, GAM and GMM — Chapter 2.
2. How to interpret formulas — in particular, the meaning and use of subscripts, parentheses, and brackets. These ideas are implicit in Chapter 2 — see Chapter 3 Summary and Self-Test, Study Guide; Readings.

Examples: The formula P_4O_{10} refers to a molecule in which there are *four atoms of phosphorus combined with ten atoms of oxygen*. As you will see in this chapter, the formula itself may refer to Avogadro's number of molecules, depending on the context in which the formula is used. But in any context, the atom ratio is four to ten in the molecule. The formula $Ca(NO_3)_2$ refers to a unit of structure in which there are *one Ca atom, two N atoms, and six O atoms*, or to Avogadro's number of such formula units. (The parentheses set off atoms which together may act as a unit; the subscript 2 then refers to two such groups of atoms.)

Math

1. How to work with conversion factors — Chapter 1.
2. How to deal with significant figures — Chapter 1.
3. How to relate gram formula masses (GAM, GMM), masses, and numbers of particles — Chapter 2.

CHAPTER SUMMARY

Stoichiometry is the awesome label usually attached to the arithmetic of chemistry. It is the quantitative application of chemical formulas and equations. The practical importance of stoichiometry is easily demonstrated — the principles established in this chapter will be applied throughout the text and in any quantitative laboratory work you do. And beyond this course: the mining engineer may want to know the amount of iron he can expect to extract from a given amount of the ore hematite, Fe_2O_3; the botanist may wish to determine the volume of oxygen liberated when a certain mass of glucose is photosynthesized; the weight-watcher may want to compare the energies stored in equal masses of carbohydrate and fat.

This chemical arithmetic is not particularly difficult or complicated; but its mastery may require considerable practice. A work-saving "mole method" of solving stoichiometric problems involves a consistent interpretation of all amounts of chemical species (atoms, ions, molecules, electrons, ...) in terms of a single unit, the *mole*.

A mole measures a unit amount of a single chemical species or substance. This means two things: a mole is a *number* of particles or formula units; a mole represents a definite *mass*, associated with this number of particles.

    1. A mole is a counting unit with exactly the same kind of meaning and uses as other, more familiar counting units, such as *dozen*. A mole is Avogadro's number of identical items; the dozen, twelve items. Where we have used the words "gram atomic mass," "gram molecular mass," and "gram formula mass" we now substitute the one word "mole."

    2. A mole is thus a variable unit of mass, its value (in grams) depending on the formula mass.

If you are baffled by the concept of mole, consider the following analogy. A "dozen identical objects" could just as well mean, simultaneously, a certain mass of objects and their number. If a penny weighs three grams, then counting out a dozen pennies is equivalent to weighing out 36 g of pennies. Any amount of pennies, specified in terms of the unit dozen, will automatically mean a certain number of pennies *and* a certain mass of pennies. A unit amount (i.e., a dozen) of dimes will contain the same number of coins as a dozen pennies but will have a different mass (about 27 g). We could conveniently count *or* weigh out any amount of identical coins in units of dozen. In particular, we could be sure that, say, 72 g of pennies and 54 g of dimes contained equal numbers of coins.

The mole concept thus provides the chemist with the means of obtaining any desired number of formula units by making the convenient measurement of mass. (For example, satisfy yourself that 71 g of chlorine, Cl_2, contain twice as many molecules as does 1.01 g of hydrogen, H_2.)

The mole method of solving problems might be outlined as follows:
    1. Write the balanced chemical equation for the reaction.
    2. Use the coefficients of the equation as conversion factors for relating the numbers of moles of given and desired species.
    3. Convert the given amounts to moles by using the appropriate gram formula masses (i.e., grams per mole).
    4. Convert from moles of desired species to mass, if needed, by using gram formula masses, or to numbers of particles by using Avogadro's number (i.e., particles or formula units per mole), thus expressing the result in the units called for.

28 • 3–Chemical Formulas and Equations

More often than not, the units of the desired quantity will indicate the kind of calculation you need perform. Suppose, for example, that you are given the mass of a reactant and are asked to find the mass of a certain product in a reaction. The series of conversions must take you from "grams reactant" through a relation of the numbers of moles of reactant and product, finally to the "grams product." The conversions called for in applying the mole method would look like this:

$$\text{g product} = \text{g reactant} \times \underbrace{\frac{1 \text{ mol reactant}}{\text{GFM reactant}}}_{\text{I}} \times \underbrace{\frac{\text{no. moles product}}{\text{no. moles reactant}}}_{\text{II}} \times \underbrace{\frac{\text{GFM product}}{1 \text{ mol product}}}_{\text{III}}$$

Note that conversion factor I converts grams to moles; conversion factor II is the ratio of the coefficients gotten from the balanced equation; conversion factor III gives the desired units, converting from moles to grams.

BASIC SKILLS

1. Relate the number of moles of a species to the number of grams or the number of particles.

As pointed out in the text discussion and above, the mole can represent either a specific number of particles (6.022×10^{23}) or a specific mass in grams (one gram formula mass). When used in the latter context, the formula of the substance must be specified. Thus:

$$1 \text{ mol Na}_2\text{CO}_3 = 2(22.99 \text{ g}) + 12.01 \text{ g} + 3(16.00 \text{ g}) = 105.99 \text{ g}$$

$$1 \text{ mol CCl}_4 = 12.01 \text{ g} + 4(35.45 \text{ g}) = 153.81 \text{ g}$$

Conversions between moles and grams are illustrated in Example 3.5; Problem 3.3 is entirely analogous. Both meanings of the mole are involved in Problems 3.16, 3.17, and 3.34. The fact that the mole can represent Avogadro's number of any kind of particles (atoms, molecules, or ice cubes) is emphasized in Problem 3.35.

2. Given the formula of a compound, calculate the percentages by mass of the elements.

Basic Skills • 29

This skill is illustrated in Example 3.6. A similar calculation is involved in Problems 3.4, 3.8, and 3.26. Also, Problems 3.9 and 3.27 are readily solved if you first determine the percentages of tin and fluorine in SnF_2 and the percentage of carbon in aspirin.

3. Determine the simplest formula of a compound, given the percentages by mass of the elements or analytical data from which these percentages can be calculated.

A simple illustration of the determination of a simplest formula from percentage composition is given in Example 3.1. Example 3.2 shows how the percentages of the elements can be calculated from data obtained by chemical analysis. Finally, Example 3.3 shows a two-step calculation, first of percentage composition and then of a simplest formula.

The calculation of simplest formula from % composition is illustrated in Problems 3.1, 3.10, and 3.28. In Problems 3.11, 3.12, 3.14, 3.29, 3.30, and 3.32, analytical data are given from which you should be able to obtain simplest formulas, either by working through % composition, as in Example 3.3 of the text, or more directly, as explained directly following that example.

4. Given the simplest formula of a substance and an approximate value for its molecular mass, obtain its molecular formula.

This skill is illustrated in Example 3.4 and in Problems 3.13 and 3.31.

5. Write and balance simple equations.

This skill is described in Section 3.5 of the text where the balanced equation is derived for the reaction of N_2H_4 with N_2O_4. A somewhat simpler example, dealing with a reaction in water solution, is presented here.

Write a balanced equation to represent the reaction that takes place when water solutions containing silver ions, Ag^+, and sulfide ions, S^{2-}, are mixed to give a precipitate of silver sulfide. _____

In order to write the balanced equation, we must first deduce the formula of the solid product. To achieve electrical neutrality, two Ag^+ ions are required to balance one S^{2-} ion. The simplest formula of silver sulfide must then be Ag_2S. The balanced equation follows directly:

$$2\,Ag^+(aq) + S^{2-}(aq) \rightarrow Ag_2S(s)$$

30 • 3–Chemical Formulas and Equations

The symbol (aq) is used to indicate that the reactants, Ag^+ and S^{2-} ions, come from water solutions. These might, for example, be water solutions of silver nitrate, $AgNO_3$, and sodium sulfide, Na_2S.

Later in the course we shall have a great deal more to say about writing and balancing equations. Equations of the type just written, often referred to as net ionic equations, are discussed in some detail in Chapter 18 of the text. A systematic method of balancing rather complex oxidation-reduction equations is described in Chapter 22. With a little bit of luck, you should be able to use the simple approach outlined in this chapter to work problems such as 3.5, 3.18, and 3.36. At most, you will be expected to know the formulas and ordinary physical states of such common substances as carbon dioxide, water, and oxygen.

6. **Given a balanced equation, relate the numbers of moles, grams, or particles of two substances taking part in the reaction.**

The conversions required here are illustrated in some detail in Example 3.7. Notice the consistent use of conversion factors here, as in previous chapters. Experience over many years convinces us that this approach is the most general and reliable way of analyzing a wide variety of problems in general chemistry. If you are still using the "ratio and proportion" method, it's high time you switched. Most of the problems that you will be assigned for homework or asked to solve on exams simply cannot be worked by a rote approach. Besides, it will help you to understand the examples worked here and in the text if we are using the same language.

Problems 3.6, 3.19, and 3.37 illustrate this skill in perhaps its simplest form. Problems 3.20 to 3.23 and 3.38 to 3.41 use the same principle but involve one or more additional conversions. In Problem 3.21, for example, you might first calculate the number of moles of H_2SO_4 required and then apply the balanced equation to relate moles of H_2SO_4 to grams of FeS.

7. **Given the masses of all reactants, determine which is the limiting reagent, and calculate the theoretical yield of product.**

This skill is described in some detail in the text discussion (Section 3.7) and in Example 3.8. Perhaps a nonchemical example will further illustrate the reasoning involved.

A bartender has available a quart bottle of gin and another quart bottle of vermouth. How many quarts of martinis (4 parts gin to 1 part vermouth) can he prepare?_____

Clearly, the "limiting reagent" is the gin; to use all the vermouth would require that four quarts of gin be available rather than one. If the bartender uses all the gin, he will need only $\frac{1}{4}$ quart of vermouth and will get $\frac{5}{4}$ quart of martinis. The "equation" is

$$1 \text{ qt gin} + \frac{1}{4} \text{ qt vermouth} \rightarrow \frac{5}{4} \text{ qt martinis}$$

This skill is applied in a more sober way to work Problems 3.7, 3.24, and 3.42.

8. Relate the actual yield of product in a reaction to the theoretical yield and the % yield.

These three quantities are related by the defining equation:

$$\% \text{ yield} = \frac{\text{actual yield}}{\text{theoretical yield}} \times 100\%$$

Clearly if any two of the three quantities are known, the third can be calculated. See the discussion immediately following Example 3.8 in the text, and the following example.

A student carrying out a reaction gets 12.0 g of product. This represents 52% of the yield he should have obtained if the limiting reagent were completely consumed. What was the theoretical yield? _____
Solving the above equation for theoretical yield:

$$\text{theoretical yield} = \text{actual yield} \times \frac{100\%}{\% \text{ yield}}$$

$$= 12.0 \text{ g} \times \frac{100}{52} = 23 \text{ g}$$

NOTES ON WRITING CHEMICAL EQUATIONS

It may appear that you are being asked to write and understand chemical equations of lots of reactions when you don't yet know any chemistry. How, for example, do you know which substances are gases, which are solids? Or which substance is composed of molecules and which is not — and so what kind of formula to use? Or what conditions are required for the reaction to proceed as indicated by the equation? Our main concern at this point is with establishing a systematic approach to solving stoichiometric problems, and this does require at least some writing of balanced chemical equations. *The descriptive chemistry will be learned as you go along.*

So, what information do you need to be given and what do you need to find on your own, and where, so as to write equations? The following suggestions are to assist you in answering these and other questions. Further practice and suggestions can be found in Chapters 18 to 23, as well as in the Readings.

An equation represents, both qualitatively and quantitatively, observed changes in matter in terms of the rearrangements of atoms, ions, and molecules. An equation is "balanced" when it reflects the observed mass relationships between reactants and products, including the conservation of matter. Balancing requires that the same kinds and numbers of atoms appear in the products as started out in the reactants. To write a chemical equation ("balanced" is generally taken for granted):

1. You need to know, or be given, the reactants and products actually observed under the given reaction conditions.

> At first, you will be given the names and formulas; gradually, you will acquire criteria for deciding what the substances are. Example: CO_2 and H_2O are the usual products in the "combustion" reactions of carbon- and hydrogen-containing compounds with oxygen and air.

> You will need to use the molecular formula when a substance is known to be molecular – otherwise, the simplest formula. The maximum information possible should be incorporated in an equation.

2. Show the physical states of all reactants and products.

> These must be the states you would observe under the given reaction conditions. Example: If water is a product in a reaction occurring at room temperature, you would expect it to be a liquid.

3. Follow the various conventions such as writing the reactants (the substances consumed or disappearing during a reaction) to the left, products to the right, separated by an arrow ($\rightarrow$), and using the simplest ratio of whole numbers of formula units needed to balance the equation. Balancing requires choosing the relative numbers of formula units so as to conserve atoms – not rewriting the formulas themselves! Example:

> The formation of liquid water from hydrogen and oxygen gases, under almost all conditions, can be represented by the equation $2 H_2(g) + O_2(g) \rightarrow 2 H_2O(l)$ but *not* by the "equation" $H_2(g) + O(g) \rightarrow H_2O(l)$, where the formula of oxygen has been manipulated so as to achieve a balance. (The formulas must correspond to reality; oxygen is known to be diatomic.)

Finally, be aware of some of the limitations of chemical formulas and equations. Beyond specifying the physical state, a formula in itself, or an equation, does not say anything directly about the conditions needed for the reaction to occur, or indeed whether the reaction can occur. An equation says nothing about reaction speed, nothing about the extent to which a given reactant is consumed and, usually, nothing about how the reaction actually occurs among the individual atoms and molecules. All of these questions need to be answered by observations that are not represented by an equation.

SELF-TEST

True or False

1. Since three-fourths of the atoms in a sample of $ScCl_3$ are ( ) chlorine atoms, 75% of the mass of the sample is due to chlorine.

2. The % by mass of iron is greater in $FeCl_2$ than in $FeCl_3$. ( )

3. The molecular formula for hydrogen peroxide, H_2O_2, ( ) indicates that one *molecule* of hydrogen (H_2) is bonded to one *molecule* of oxygen (O_2).

4. In fifty grams of calcium carbonate, $CaCO_3$ (FM = 100), ( ) there are 3×10^{23} Ca atoms, 0.5 mol of C, and $\frac{3}{2}$ GAM of oxygen.

5. The calculation of theoretical yield is based on the ( ) assumption that all of the limiting reagent is consumed according to the equation for the reaction.

6. All of the following may be reasons why the actual product ( ) yield in a reaction is usually less than 100%: separation and purification result in losses; competing reactions form other products instead; the reaction hasn't stopped yet; not all the reactants are converted to desired product, even when the reaction has ceased.

7. The chemical mole is defined as being Avogadro's number ( ) of formula units (molecules, ions, atoms, electrons ...).

8. In an ordinary chemical reaction, the number of moles of ( ) reactants always equals the number of moles of products.

9. In each mole of $(NH_4)_2CO_3$ there are three moles of ions. ( )

34 • 3–Chemical Formulas and Equations

10. From chemical analysis one finds that 1.801 mol of Cl ( )
combine with 0.9017 mol of Ca. With the atom ratio of Cl to Ca
being 1.801/0.9017 = 1.997, one should give the formula of calcium
chloride as $CaCl_{1.997}$ and not $CaCl_2$.

Multiple Choice

11. In analyzing a compound for carbon, the compound is ( )
burned in air and the masses of products determined. What
assumption do we make?
    (a) all the oxygen in the air is converted to water, H_2O
    (b) all the carbon is converted to carbon dioxide, CO_2
    (c) equal numbers of moles of CO_2 and H_2O are produced
    (d) all the oxygen in the H_2O produced comes from the compound

12. The simplest formula of a substance always shows ( )
    (a) the element(s) present and the simplest ratio of whole numbers of atoms
    (b) the actual numbers of atoms combined in a molecule of the substance
    (c) the number of molecules in a sample of the substance
    (d) the gram molecular mass of the substance

13. What *minimum* information would be sufficient for determining the simplest formula for a compound? ( )
    (a) the elements present in the compound
    (b) the elements in the compound and their atomic masses
    (c) the elements in the compound, their atomic masses, and their combining masses in a sample of the compound
    (d) the elements in the compound, their atomic masses, and their combining masses and mass percentages in a sample of the compound

14. In order to determine the composition of a substance, one might ( )
    (a) determine its physical properties and compare them to those of known substances
    (b) convert the substance to one or more substances of known composition
    (c) compare the products for each of several reactions to those gotten when similarly reacting other, known substances
    (d) all of the above

15. Information given by the molecular formula for hydrazine, ( )
N$_2$H$_4$, includes all of the following except:
    (a) hydrazine could just as well be represented by the formula NH$_2$
    (b) the per cent by mass that is nitrogen is (28.0/32.0) × 100%
    (c) one molecule of hydrazine contains six atoms
    (d) one mole of hydrazine weighs 32.0 g

16. The formula for calcium carbonate (CaCO$_3$, FM = 100) ( )
generally represents all of the following except:
    (a) one formula unit ("molecule") of calcium carbonate
    (b) N formula units of calcium carbonate
    (c) one gram of calcium carbonate
    (d) one hundred grams of calcium carbonate

17. A compound with the simplest formula C$_2$H$_5$O has a ( )
molecular mass of 90. The molecular formula for the compound is
    (a) C$_3$H$_6$O$_3$ (b) C$_4$H$_{26}$O
    (c) C$_4$H$_{10}$O$_2$ (d) C$_5$H$_{14}$O

18. Which contains the largest number of molecules? ( )
    (a) 1.0 g CH$_4$ (MM = 16)   (b) 1.0 g H$_2$O (MM = 18)
    (c) 1.0 g HNO$_3$ (MM = 63)   (d) 1.0 g N$_2$O$_4$ (MM = 92)

19. About how much oxygen is there in exactly one mole of ( )
baking soda, NaHCO$_3$?
    (a) 16 g   (b) 24 g
    (c) 48 g   (d) 96 g

20. It is estimated that there are 1 × 10^{21} kg of water in all the ( )
oceans. How many moles of water is this?
    (a) $\dfrac{1 \times 10^{21}}{18}$ (b) $\dfrac{1 \times 10^{21} \times 10^3}{18}$
    (c) 1 × 10^{21} × 18 × 10^3 (d) $\dfrac{1 \times 10^{21}}{6 \times 10^{23}}$

21. The molecular mass of a protein that causes food poisoning ( )
is about 900 000. The approximate mass of one molecule of this protein is
    (a) 2 × 10^{-18} g   (b) 1 × 10^{-6} g
    (c) 9 × 10^5 g   (d) some other number

22. A balanced chemical equation shows ( )
    (a) the mole ratio in which substances react
    (b) the direction a chemical system will move in and the extent or yield of the reaction

36 • 3–Chemical Formulas and Equations

(c) the speed with which the reaction proceeds
(d) the individual molecular steps by which the reaction occurs

23. To write a chemical equation for a given reaction, you ( ) would need to know at least
   (a) the masses of the reactants and products in the given reaction
   (b) the mole ratios of all reactants and products in the reaction
   (c) the formulas of all reactants and products in the reaction
   (d) all of the above

24. Ammonia (NH_3) and oxygen (O_2) can be made to react to ( ) form only nitrogen and water. The number of moles of oxygen consumed for each mole of nitrogen formed is
   (a) 1.5   (b) 0.67
   (c) 3.0   (d) 2.0

25. Which equation most completely represents the following ( ) reaction? On being heated, gaseous ammonia (NH_3) decomposes to form gaseous nitrogen (N_2) and hydrogen (H_2).
   (a) $NH_3(g) \rightarrow N_2(g) + H_2(g)$
   (b) $3 H_2(g) + N_2(g) \rightarrow 2 NH_3(g)$
   (c) $2 NH_3 \rightarrow 3 H_2 + N_2$
   (d) $2 NH_3(g) \rightarrow 3 H_2(g) + N_2(g)$

26. When balanced, the following equation for the combustion ( ) of octane has which set of coefficients?

   ___C_8H_{18}(l) + ___O_2(g) → ___CO_2(g) + ___H_2O(l)

   (a) 1, 25, 8, 18   (b) 1, 25/2, 16, 18
   (c) 1, 25, 8, 9   (d) 2, 25, 16, 18

27. If 4.5 mol of NO_2 are allowed to react with 3.0 mol of H_2O ( ) according to the equation

$$3 NO_2(g) + H_2O(l) \rightarrow 2 HNO_3(l) + NO(g)$$

the theoretical yield of HNO_3, in moles, would be
   (a) 4.5   (b) 3/2 × 4.5
   (c) 2.0   (d) 2/3 × 4.5

28. Baking soda ($NaHCO_3$) and hydrochloric acid (HCl) react ( ) according to the equation $HCO_3^-(aq) + H^+(aq) \rightarrow H_2O(l) + CO_2(g)$.

At least how many moles of NaHCO₃ are required for the formation of 2.5 mol of CO₂?
 (a) 1.0   (b) 2.5
 (c) 5.0   (d) some other number

29. How many grams of baking soda (FM = 84.0) would be ( ) consumed in forming 10.0 g of carbon dioxide (MM = 44) according to the equation given in (28)?

 (a) $\dfrac{10.0}{44}$ (b) $\dfrac{44 \times 84.0}{10.0}$

 (c) $\dfrac{10.0 \times 84.0}{44}$ (d) $\dfrac{10.0 \times 61.0}{44}$

30. When balanced using the smallest whole numbers possible, ( ) the coefficient for Cl⁻ is

$$\underline{\quad}Cl_2(aq) + \underline{\quad}OH^-(aq)$$
$$\rightarrow \underline{\quad}Cl^-(aq) + \underline{\quad}ClO_3^-(aq) + \underline{\quad}H_2O(l)$$

 (a) 1   (b) 3
 (c) 4   (d) 5

31. When a certain compound of nitrogen and hydrogen reacts ( ) with oxygen, nitric oxide (NO) and water are formed. Under reaction conditions, all four substances are gases and are found to react in a volume ratio of 4:5:4:6, respectively. What is the simplest formula for the nitrogen-hydrogen compound?
 (a) NH₂   (b) N₂H₄
 (c) HN₃   (d) NH₃

32. In carrying out the reaction   ( )

$$2\ NaHCO_3(s) \rightarrow Na_2CO_3(s) + H_2O(g) + CO_2(g)$$

a student obtains an 80% yield. How many moles of NaHCO₃ must she have started with if her yield was 1.6 mol of Na₂CO₃?
 (a) 4.0   (b) 3.2
 (c) 2.6   (d) 2.0

33. For the reaction $C_3H_8(g) + 5\ O_2(g) \rightarrow 3\ CO_2(g) + 4\ H_2O(l)$, ( ) it is found in a certain container that 2 mol of O₂(g) are left unreacted when an initial 5 mol of O₂(g) had been used. How many moles of C₃H₈(g) have reacted?
 (a) 5/3   (b) 1
 (c) 3/5   (d) 2/5

34. The maximum amount of silver one might hope to recover ( ) from a kilogram of waste silver chloride that is only 85% AgCl would

3–Chemical Formulas and Equations

be, in kilograms (AM = 108 for Ag; FM = 143 for AgCl)
    (a) 108/143   (b) 0.85 × 108/143
    (c) 0.85 × 143/108   (d) 0.85 × 108/143 × 1000

35. When an excess of one reactant is used   ( )
    (a) more product may form than with no excess
    (b) reaction may proceed at a faster rate
    (c) less limiting reagent may remain unconsumed
    (d) all of the above

Problems

36. A 0.600 g sample of pure tin, Sn, is reacted completely with gaseous fluorine, F_2, to form 0.984 g of solid tin fluoride.
    (a) What is the simplest formula of tin fluoride? (AM Sn = 119, F = 19.0)
    (b) Write the balanced equation for the synthesis of tin fluoride.

37. A compound was shown to contain only cesium (AM = 133) and chlorine (AM = 35.5). When 2.52 g of this substance was dissolved in water and treated with excess Ag^+, 2.150 g of silver chloride, AgCl (FM = 143) was formed. Determine
    (a) the number of grams of Cl in the original sample.
    (b) the number of grams of Cs in the original sample.
    (c) the simplest formula of the compound.

38. When acetylene gas, C_2H_2, burns in air, the products are $CO_2(g)$ and $H_2O(l)$.
    (a) Write a balanced equation for the reaction.
    (b) If one mole of C_2H_2 is burned with 64.0 g of O_2, which is the limiting reagent?
    (c) What is the theoretical yield in grams of CO_2? (AM C = 12.0, H = 1.0, O = 16.0)

*39. Analysis of a gaseous mixture of NO and N_2O shows it to contain 52% by mass nitrogen. What is the % by mass of NO in the mixture?

*40. Write and execute a computer program to find the simplest formulas for the following five oxides of nitrogen:

	(1)	(2)	(3)	(4)	(5)
Mass % N	63.6	46.7	36.9	30.4	26.0
Mass % O	36.4	53.3	63.2	69.5	74.0

SELF-TEST ANSWERS

1. F (Atom for atom, Sc is heavier than Cl; actual % Cl = 70.)
2. T
3. F (The formula denotes only that there are two H atoms and two O atoms in a molecule of hydrogen peroxide.)
4. T
5. T
6. T (These answers are based on material of Chapters 1, 15, and 16 as well as on actual laboratory experience.)
7. T
8. F
9. T (From repeated exposure, as well as more systematic treatment in Chapters 8 and 18, you will begin to recognize common units of structure, such as the ions NH_4^+ and CO_3^{2-}.)
10. F (The data is from Example 3.3 in the text. Besides the possibility that the data may not be this good in an actual analysis, what reasons can you offer for rounding-off?)
11. b
12. a
13. c (Note that combining masses and mass percentages provide the same information. Also see Question 27 in Chapter 2 Self-Test.)
14. d (Any one or all of these might work.)
15. a (NH_2 would not give the correct number of atoms in a molecule of hydrazine, as indicated by the molecular formula.)
16. c
17. c (All of these formulas correspond to a MM of 90 but only (c) has the atom ratio given also by the simplest formula.)
18. a (The largest fraction of a mole, 1.0/16.)
19. c
20. b (Don't forget to convert to grams!)
21. a (Mass per molecule = 9×10^5 g/6×10^{23} molecules.)
22. a
23. c (Note that the information in (a) and (b) can be gotten from this choice, once the equation is balanced.)
24. a (You need to know that nitrogen is N_2 and to balance the equation.)
25. d (Physical states should be shown.)
26. d
27. d (Note that NO_2 is the limiting reagent; 4.5 mol $NO_2 \times \dfrac{2 \text{ mol } HNO_3}{3 \text{ mol } NO_2}$.)

40 • 3–Chemical Formulas and Equations

28. b (Each mole of HCO_3^- reacting requires the use of one mole of $NaHCO_3$.)
29. c (To check, attach appropriate units.)
30. d (Difficult? A systematic approach to balancing such a redox equation is given in Chapter 22. The respective coefficients are: 3, 6, 5, 1, 3.)
31. d (Recall Avogadro's hypothesis — Chapter 2: Volume ratios are equivalent to mole ratios for gases, all reacting under a given set of conditions.) The equation:

$$4\ NH_3(g) + 5\ O_2(g) \rightarrow 4\ NO(g) + 6\ H_2O(g)$$

32. a (For 80% yield, every 2 mol $NaHCO_3$ gives 0.8 mol Na_2CO_3.)
33. c (Three moles O_2 react: moles C_3H_8 = 3 mol O_2 × $\dfrac{1\text{ mol }C_3H_8}{5\text{ mol }O_2}$.)
34. b
35. d (Choices (a) and (c) are discussed in Chapter 15; (b), in Chapter 16.)
36. (a) 0.600 g Sn × $\dfrac{1\text{ GAM Sn}}{119\text{ g Sn}}$ = 0.005 04 mol Sn; 0.384 g F × $\dfrac{1\text{ GAM F}}{19.0\text{ g F}}$ = 0.0202 mol F; $\dfrac{0.0202}{0.005\ 04}$ = 4; SnF_4
    (b) $Sn(s) + 2\ F_2(g) \rightarrow SnF_4(s)$
37. (a) 2.150 g AgCl × $\dfrac{35.5\text{ g Cl}}{143\text{ g AgCl}}$ = 0.534 g Cl
    (b) 2.52 g − 0.534 g = 1.99 g
    (c) 1.99 g Cs × $\dfrac{1\text{ GAM Cs}}{133\text{ g Cs}}$ = 0.0150 mol Cs
         0.534 g Cl × $\dfrac{1\text{ GAM Cl}}{35.5\text{ g Cl}}$ = 0.0150 mol Cl; CsCl
38. (a) $2\ C_2H_2(g) + 5\ O_2(g) \rightarrow 4\ CO_2(g) + 2\ H_2O(l)$
    (b) 1.00 mol C_2H_2, 2.00 mol O_2; O_2 is limiting reagent
    (c) 2.00 mol O_2 × $\dfrac{4\text{ mol }CO_2}{5\text{ mol }O_2}$ × $\dfrac{44.0\text{ g }CO_2}{1\text{ mol }CO_2}$ = 70.4 g CO_2

*39. In 100 g mixture, there are X g NO:

$$52\text{ g N} = \dfrac{14\text{ g N}}{30\text{ g NO}}(X\text{ g NO}) + \dfrac{28\text{ g N}}{44\text{ g N}_2\text{O}}(100 - X\text{ g N}_2\text{O})$$

X = 71 g;   71% NO

*40. N_2O, NO, N_2O_3, NO_2, and N_2O_5

Of course, these can be calculated by hand. More interesting would be to write the program like that discussed in Chapter 7 of Soltzberg. (See Readings.)

SELECTED READINGS

Problem solving requires practice. For a selection of exercises, work as many of the problems in the text as you can. Also, consider one or more of the problem or programmed manuals listed in the Preface to this guide. And, particularly for balancing equations, consider:

Copley, G. N., Linear Algebra of Chemical Formulas and Equations, *Chemistry* (October 1968), pp. 22–27.

Greene, D. G. S., An Algebraic Method for Balancing Chemical Equations, *Chemistry* (March 1975), pp. 19–21.

Kieffer, W. F., *The Mole Concept in Chemistry*, New York, D. Van Nostrand, 1973.

Strong, L. E., Balancing Chemical Equations, *Chemistry* (January 1974), pp. 13–15.

The use of computers in chemistry is discussed in:

Soltzberg, L., *BASIC and Chemistry*, Boston, Houghton-Mifflin, 1975.

4
THERMOCHEMISTRY

QUESTIONS TO GUIDE YOUR STUDY

1. How many chemical reactions going on in the world around us can you think of in which the energy change plays an important role? (What "drives" an automobile engine? A mountain climber?)
2. What kinds of energy may be transferred during chemical reactions?
3. What is the source of the energy involved in a reaction? What happens to it? Can this energy be rationalized in terms of what happens to the atoms?
4. Is the energy change of a reaction quantitatively related to the reacting masses? (Can you cite a simple example?)
5. How is information concerning energy change expressed within the chemical equation? Does your interpretation of the energy change depend on how you write the equation for a reaction?
6. Does the energy change depend on the conditions under which a reaction is carried out? If so, how?
7. How do you measure the energy transfer for any given reaction? Can it be calculated for a reaction, without ever actually carrying out that particular reaction?
8. What do thermochemical principles allow you to say about practical problems, such as the relative merits of two different fuels?
9. What are some of the fundamental problems in supplying the energy needed by modern society today and tomorrow? Are there known limitations or perhaps untapped possibilities the chemist can describe?
10. What is thermal pollution? What can be done to minimize it?
11.

12.

YOU WILL NEED TO KNOW

Concepts

1. A general notion as to the meaning of energy and the means by which it may be transferred from one object to another — see the Summary, as well as the Readings.

2. How to add two or more chemical equations to give a single equation. In this regard chemical equations are treated as though they were algebraic equations — see the discussion of Hess's Law in Chapter 4.

Math

1. How to work problems in stoichiometry, and therefore how to write simple balanced equations — Chapter 3.

CHAPTER SUMMARY

In every chemical reaction we observe that the mass of the products is equal to the mass of the reactants. We account for mass conservation in writing equations that are "balanced." We likewise have a bookkeeping system for energy, since it too can always be accounted for. In a "thermochemical" equation we note how much energy is involved and show whether it is entering or leaving the system (the reaction mixture). The stoichiometry is as before — the equation is taken to represent amounts of substances in units of mole, with the accompanying energy change having the numerical value written into the equation or alongside it.

The usual way in which we measure the energies associated with chemical reactions is to measure a property of the surroundings, such as temperature change, that shows how much energy has left the reaction mixture and entered the surroundings, or has left the surroundings and entered the system. In all reactions, the energy is thought of as either being stored in the atoms and molecules of the system (the energy of the system increases, the sign of the energy change is positive) or being taken out of storage and passed on to the surroundings (where the energy is stored, again, in the constituent particles). What happens during a chemical reaction if the system is insulated from surrounding matter? Example: If hydrogen and oxygen explosively react to form water inside a sealed and insulated "bomb," the energy that would otherwise have been given off to the surroundings ends up being stored in the products, raising their temperature and increasing the vigor of their molecular motions.

The transfer of energy of most concern to the chemist is that called heat. Since most reactions occur open to the atmosphere, as in test tubes, we are interested in this thermal energy transferred at constant pressure. This is called the enthalpy change, ΔH. (For most purposes, even if the pressure does change during a reaction, as usually happens in a bomb calorimeter, the heat flow is essentially equal to ΔH.)

How do we associate energy with a molecule? In what ways can energy be stored? From where does it come? By the very fact that molecules are constantly in motion, they possess energy: the more energy, the more vigorous the motion. (What kind of evidence is there for molecular motion?) More about kinetic energy will be discussed in the next chapter. And several kinds of motion may be possible for a molecule, atom, or ion: vibrating, rotating, as well as just moving headlong. And there's the stored "chemical energy" associated with the bonds between atoms, a result of mutual attractions between electrons and nuclei. (Chapters 6, 8.) Differences from one molecule to the next kind of molecule in the types and numbers of these energy-storing mechanisms give rise to differences in the energy changes observed. For example, a lot more energy is normally stored in a chemical bond between the atoms in a molecule than in the motions of the molecule itself. Or again, much more energy is involved in nuclear changes than in the ordinary bond-making and bond-breaking of chemical reactions (see Chapter 24).

The fact that we can keep track of energy transfers, that energy always seems to be conserved in chemical reactions, means that we acquire the power of prediction. We can often predict the energy effect associated with a given reaction, even without first carrying out the reaction. And this is really the chief concern of this chapter — that you be able to calculate the amount of heat produced by or required for any given reaction, whether at constant pressure or volume.

But perhaps there is a more important concern for the material discussed in this chapter: that of objectively looking at our energy resources on this planet, their uses and abuses, the laws and limitations energy changes seem to follow. In essence, the First Law of Thermodynamics says that energy is not really consumed at all — it can only be changed in form; for example, from chemical energy into heat. (The Second Law might be taken to mean that energy cannot be completely recycled; more and more, energy becomes unavailable as wasteful heat in the surroundings. In Chapter 14, we will see a rationale for the "inevitable" waste of energy in terms of the behavior of particles.) How, then, do we make the best of what we've got?

4–Thermochemistry

BASIC SKILLS

1. **Use a thermochemical equation to relate heat flow in a reaction to amounts of products or reactants.**

This skill is illustrated in Example 4.1 and in the following example.

For the reaction $CH_4(g) + 2\ O_2(g) \rightarrow CO_2(g) + 2\ H_2O(l)$, $\Delta H = -890$ kJ. How many grams of CH_4 must be burned to evolve 1.00 kJ of heat?

Here, as always in working problems dealing with balanced equations, we follow the conversion factor approach. The thermochemical equation gives us the conversion factor we need, that between grams of methane and amount of heat evolved:

$$1 \text{ mol } CH_4 = 16.0 \text{ g } CH_4 \mathrel{\widehat{=}} -890 \text{ kJ}$$

To evolve 1.00 kJ of heat, we need:

$$-1.00 \text{ kJ} \times \frac{16.0 \text{ g } CH_4}{-890 \text{ kJ}} = 0.0180 \text{ g } CH_4$$

Problems 4.1, 4.8, and 4.24 are solved in exactly the same way as these examples. Problem 4.9, 4.10, 4.25, and 4.26 are similar, but a little more advanced.

2. **Calculate ΔH for a reaction from heats of formation of compounds (Table 4.1).**

This is readily accomplished by applying Equation 4.8. Be careful about signs; since most heats of formation are negative, the products will ordinarily make a negative contribution to ΔH, the reactants a positive contribution. Example 4.2 illustrates this skill in its simplest form. Example 4.3 shows how Equation 4.8 can be used to obtain the heat of formation of a compound, knowing ΔH and the heats of formation of all other substances taking part in the reaction.

To gain practice in this skill, try Problem 4.2, which is entirely analogous to Example 4.2. Problems 4.12 and 4.28a combine this skill with Skill 1 above. Finally, Problems 4.13 and 4.29 are entirely analogous to Example 4.3.

Basic Skills • 47

3. **Calculate ΔH for a reaction from bond energies (Table 4.2).**

See Example 4.4 and Problems 4.14 and 4.30. Note that to use this skill you must:

a. know what bonds are present in reactant and product molecules. At this stage, prior to a discussion of chemical bonding (Chapter 8), the nature of the bonds will ordinarily be indicated in the statement of the problem.

b. apply Hess's Law (see discussion in text). We make extensive use of this law in this and succeeding chapters. Problems 4.11 and 4.27 give you practice in using it.

4. **Use calorimetric data to obtain:**

a. **the specific heat of a substance.** This skill is illustrated in Example 4.5 and the discussion preceding it. The basic principle here is that the heat flow for the substance must be equal in magnitude but opposite in sign to that of the calorimeter. In a coffee-cup calorimeter containing water:

$$Q \text{ substance} = -Q \text{ water}$$

$$(m \text{ substance})(S.H. \text{ substance})(\Delta t \text{ substance})$$
$$= (m \text{ water})(S.H. \text{ water})(\Delta t \text{ water})$$

or

$$S.H. \text{ substance} = -4.18 \frac{J}{g \cdot °C} \times \frac{m \text{ water}}{m \text{ substance}} \times \frac{\Delta t \text{ water}}{\Delta t \text{ substance}}$$

(Note that since Δt water and Δt substance will always have opposite signs, the specific heat calculated from the above equation will always be positive, as it must be.)

Problems 4.4 and 4.15 illustrate this skill in its simplest form. In Problem 4.31, the calculation is turned around; knowing the specific heats and masses of both silver and water, you are asked to calculate the final temperature.

b. **Q for a chemical reaction.** The principle here is exactly the same as in (a): the heat given off by the reaction is absorbed by the calorimeter and its contents. If a bomb calorimeter is used, we must consider both the heat absorbed by the water present and that taken up by the metal parts of the calorimeter. This consideration leads directly to Equation 4.16 in the text or, going one step further

$$Q \text{ reaction} = -\left[m \text{ water} \times 4.18 \frac{J}{g \cdot °C} + C \right] \Delta t$$

where C is the total heat capacity ($J/°C$) of the bomb.

This skill is shown in Example 4.6 and must be applied to solve Problems 4.5, 4.16, and 4.32 (note that in the latter two problems, Skill 1 is also required). Problems 4.17 and 4.33 illustrate two different approaches to determining C, using the above equation. In both cases, you must first calculate "Q reaction"; i.e., the heat flow for the process taking place in the calorimeter.

5. Use the First Law of Thermodynamics to calculate ΔE, W, or Q, given the other two quantities.

This simple calculation is discussed in Example 4.7; see Problems 4.6, 4.19, and 4.35. If you don't like the minus sign in Equation 4.18, blame it on the engineers who decided a long time ago to regard Q as positive when heat flows into a system, but took W to be negative when work "flows into" (is done upon) a system.

6. Use Equation 4.24 to relate ΔH for a reaction to ΔE.

See Example 4.8 and Problems 4.7, 4.20, and 4.36. Two notes concerning this skill, one practical, the other philosophical:

– ΔH is the quantity measured when a reaction is carried out in an open container; it is also the quantity calculated when Tables 4.1 and 4.2 are used. ΔE is the quantity measured when a reaction is carried out in a sealed container such as a bomb calorimeter. The difference between ΔH and ΔE is always small, but thermodynamicists tend to be perfectionists who worry about little things.

– Equations 4.18 and 4.24 are typical of thermodynamics: very simple equations based on a precise and somewhat subtle line of reasoning. The equations are easy to use; the only difficulty is to decide under what conditions they are applicable.

SELF-TEST

True or False

1. In carrying out a reaction in a test tube, a student observes (   ) that the test tube becomes cold. He should call the reaction exothermic.

2. *At constant pressure and temperature* means that the final (   ) pressure and temperature are the same as the initial, regardless of what happens in between.

Self-Test • 49

3. Given the thermochemical equation ( )

$$UF_6(l) \rightarrow UF_6(g), \Delta H = +30.1 \text{ kJ},$$

one can be sure that 30.1 kJ of heat must be evolved when one mole of liquid UF_6 is evaporated.

4. Another way of writing the thermochemical equation of ( ) Question (3) would be $UF_6(l) + 30.1 \text{ kJ} \rightarrow UF_6(g)$.

5. You would expect that the heat produced in the following ( ) reactions would have one and the same numerical value:

$$H_2(g) + \frac{1}{2}O_2(g) \rightarrow H_2O(l); \quad 2H_2(g) + O_2(g) \rightarrow 2H_2O(l)$$

6. The difference in enthalpy between one mole of Cl_2 and ( ) two moles of atomic chlorine, Cl, both at one atmosphere pressure and 25°C, is equal in magnitude to the bond energy for one mole of Cl_2.

7. Considering the example reaction ( )

$$C(s) + O_2(g) \rightarrow CO_2(g)$$

one can always say that the heat of combustion for any substance is the same thing as the so-called heat of formation of that substance.

8. One would expect that ΔH for the following reaction would ( ) be less negative than the molar heat of formation of liquid water:

$$2H(g) + O(g) \rightarrow H_2O(l)$$

9. If a system is carried through a series of steps, each of which ( ) involves an energy transfer into or out of the system, but finally ends up being identical in every way to the initial system, the total heat flow, Q, must be zero.

10. Reactions which tend to occur of their own accord, ( ) proceeding "naturally" or spontaneously, usually involve a decrease in enthalpy.

Multiple Choice

11. The largest amount of energy is stored in which one of the ( ) following systems?

50 • 4–Thermochemistry

(a) one mole H$_2$O(l) at 100°C
(b) one mole H$_2$O(l) at 0°C
(c) one mole H$_2$O(g) at 100°C and one atmosphere
(d) one mole H$_2$O(g) at 100°C and one atmosphere and one mole H$_2$O(l) at 100°C have the same amount of stored energy

12. The regulation of body temperature by perspiration can be ( ) accounted for, at least in part, by the fact that
    (a) liquid water has a negative heat of formation
    (b) liquid water is a good insulator
    (c) the phase change, liquid-to-gas, is endothermic
    (d) gram for gram, water absorbs and releases less heat per unit temperature change than most other substances (i.e., it has a small specific heat)

13. The molar heat of combustion of methane, CH$_4$, is reported ( ) as –890 kJ. The corresponding thermochemical equation is
    (a) C(g) + 4 H(g) → CH$_4$(g) $\Delta H = -890$ kJ
    (b) C(s) + 2 H$_2$(g) → CH$_4$(g) $\Delta H = -890$ kJ
    (c) CH$_4$(g) + $\frac{3}{2}$ O$_2$(g) → CO(g) + 2 H$_2$O(l) $\Delta H = -890$ kJ
    (d) CH$_4$(g) + 2 O$_2$(g) → CO$_2$(g) + 2 H$_2$O(l) $\Delta H = -890$ kJ

14. One can often confidently calculate the enthalpy change for ( ) a reaction before carrying out the reaction because
    (a) all heats of reaction have already been measured and tabulated
    (b) heats of formation for all known compounds have been measured
    (c) the given reaction, and its enthalpy change, may be related algebraically to other reactions already carried out
    (d) since enthalpy change depends on amount of substance reacting, one can always adjust the amount of materials so as to observe an enthalpy change of any value

15. For the reaction 2 HCl(g) → H$_2$(g) + Cl$_2$(g), $\Delta H = +185$ ( ) kJ. This means that:
    (a) if the given reaction is to be carried out at constant pressure and temperature, then the reaction mixture must be heated
    (b) the chemical bonds in the products are weaker than those in the reactants
    (c) HCl(g) has a negative heat of formation
    (d) all of the above

16. When water evaporates at constant pressure, the *sign* of the ( )
heat flow
   (a) is negative
   (b) is positive
   (c) depends on the temperature
   (d) depends on the container volume

17. Given ΔH = -601.8 kJ for the reaction Mg(s) + $\frac{1}{2}$ O$_2$(g) → ( )
MgO(s), what would you expect to happen if the reaction were allowed to proceed at constant pressure in such a way that no heat transfer could take place between the reaction mixture and the surroundings?
   (a) no reaction could occur
   (b) the temperature of the reaction mixture would increase
   (c) the temperature of the reaction mixture would decrease
   (d) insufficient information is given

18. When one gram of ammonia, NH$_3$ (MM = 17.0), is produced from N$_2$ and H$_2$ at a constant temperature (25°C) and pressure (1 atm), 2720 J are evolved. The molar heat of formation of ammonia, in kilojoules, is ( )
   (a) -2.72 (17.0)   (b) 17.0/2720
   (c) -2.72/17.0   (d) +2.72 (17.0)

19. Which one of the following reactions would you expect to ( )
be the source of the largest amount of heat?
   (a) CH$_4$(l) + 2 O$_2$(g) → CO$_2$(g) + 2 H$_2$O(g)
   (b) CH$_4$(g) + 2 O$_2$(g) → CO$_2$(g) + 2 H$_2$O(g)
   (c) CH$_4$(g) + 2 O$_2$(g) → CO$_2$(g) + 2 H$_2$O(l)
   (d) CH$_4$(g) + $\frac{3}{2}$ O$_2$(g) → CO(g) + 2 H$_2$O(l)

20. In what order would you arrange the following reactions so ( )
that the magnitude of the enthalpy change increased in that order?
   A. ^{238}U + n → ^{239}U
   B. H$_2$O(l) → H$_2$O(s)
   C. C(coal) + O$_2$(g) → CO$_2$(g)
   (a) ABC   (b) BCA
   (c) CAB   (d) ACB

21. Given the heats of combustion for diamond (-395.4 kJ) ( )
and graphite (-393.5 kJ), both composed of pure carbon, would you expect the formation of diamond from graphite to be endothermic or exothermic?

52 • 4–Thermochemistry

(a) endothermic
(b) exothermic
(c) depends on the temperature
(d) insufficient information

22. Hydrogen, H_2, has been proposed as a fuel of the future. (   )
Knowing that very little terrestrial hydrogen exists in this elemental state, you would expect that
    (a) hydrogen could play no role in the transfer of energy from one system to another
    (b) elemental hydrogen might well supplant all other fuels
    (c) energy received from another source might be stored in H_2 and later released during the combustion of the hydrogen
    (d) all of the above

23. In 1970, we obtained about 23% of our energy from coal, (   ) about 75% from oil and natural gas, about 2% from water power, and about 0.2% from nuclear processes. Which one of these percentages seems most likely to decrease by 2000 A.D.?
    (a) coal   (b) oil and natural gas
    (c) water power   (d) nuclear processes

24. The greatest potential for meeting our energy needs in the (   ) long run seems to be offered by
    (a) coal   (b) petroleum
    (c) fission   (d) fusion on the earth or sun

25. In order to determine the heat evolved when a sample is (   ) burned in a bomb calorimeter, you must know
    (a) the mass and specific heat of the water in the calorimeter
    (b) the temperature change
    (c) the heat capacity of the bomb
    (d) all the above

26. Given that ΔH = +185 kJ for the reaction 2 HCl(g) → (   ) H_2(g) + Cl_2(g), the H—Cl bond energy
    (a) is −185 kJ
    (b) is −92.5 kJ
    (c) is +92.5 kJ
    (d) cannot be determined without additional information

27. The bond energies for the halogens F_2, Cl_2, Br_2, and I_2 are (   ) 153 kJ, 243 kJ, 193 kJ, and 151 kJ, respectively. Of these, the strongest bond is found in
    (a) F_2 (b) Cl_2
    (c) Br_2 (d) I_2

28. The sign of ΔH for the reaction $X_2(g) \rightarrow 2 X(g)$, where X is ( ) an atom,
    (a) is negative   (b) is positive
    (c) ΔH = 0   (d) depends on the identity of X

29. Consider the reaction $2 H(g) + O(g) \rightarrow H_2O(l)$. For this ( ) reaction, ΔH is the same as
    (a) $\Delta H_f\ H_2O(l)$ (b) $\Delta H_f\ H_2O(g)$
    (c) $-2\epsilon\ (O-H)$ (d) none of these

30. To show that in ammonia the nitrogen atom is bonded to ( ) each of three hydrogen atoms with no other bonds present, one would draw
    (a) NH_3 (b) N—H—H—H
    (c) H—N—H—H   (d) H—N—H
                                                        |
                                                        H

31. Which property of a mole of a pure substance depends on ( ) the pressure and temperature?
    (a) H   (b) E
    (c) V   (d) all of these

32. The best experimental values of thermochemical data are ( ) obtained through measurements of
    (a) ΔH   (b) Q_P
    (c) Q_V (d) PΔV

33. If 20 g of $H_2O(l)$ originally at 25°C are heated to 35°C, the ( ) heat absorbed by the $H_2O(l)$ is about
    (a) +200 J   (b) +600 J
    (c) +800 J   (d) -800 J

34. A system consisting of one mole of gaseous water at 300°C ( ) and one atmosphere is carried through the following process:
    Step 1: 225 J of heat are supplied to the system at constant volume
    Step 2: 225 J of work are done on the system; 225 J of heat are absorbed by the system
    Step 3: the system is allowed to expand until the pressure drops to one atmosphere; heat is removed from the system until the temperature is again 300°C
What is the value of ΔE, in joules, for the overall process?
    (a) insufficient data given   (b) 675
    (c) 225   (d) zero

35. For the reaction $H_2O(l) \rightarrow H_2O(g)$ at 298 K, 1 atm, ΔH is ( ) more positive than ΔE by 2.5 kJ. This quantity of energy can be considered to be

54 • 4–Thermochemistry

(a) the heat flow required to maintain a constant temperature
(b) the work done in pushing back the atmosphere
(c) the difference in the H—O bond energy in $H_2O(l)$ compared to $H_2O(g)$
(d) the value of ΔH itself

Problems

36. Consider the reaction $C_3H_8(g) + 5\ O_2(g) \rightarrow 3\ CO_2(g) + 4\ H_2O(l)$. Referring to the heats of formation below, calculate ΔH for the combustion of 1.40 g of $C_3H_8(g)$ in excess air (AM C = 12.0, H = 1.0).

	ΔH_f(kJ/mol)		ΔH_f(kJ/mol)
CO(g)	−110.5	H_2O(g)	−241.8
CO_2(g)	−393.5	H_2O(l)	−285.8
C_3H_8(g)	−103.8		

37. For the reaction $2\ C_2H_2(g) + 5\ O_2(g) \rightarrow 4\ CO_2(g) + 2\ H_2O(l)$, $\Delta H = -2599.9$ kJ. The heats of formation of $CO_2(g)$ and $H_2O(l)$ are −393.5 kJ/mol and −285.8 kJ/mol, respectively. Calculate
    (a) the heat of formation of acetylene, C_2H_2 (kJ/mol)
    (b) the final temperature reached if the heat evolved in the combustion of two moles of acetylene is absorbed by 50.0 kg of water, originally at 20.0°C

38. When the reaction $2\ KClO_3(s) \rightarrow 2\ KCl(s) + 3\ O_2(g)$ is carried out in a bomb calorimeter, 89.5 kJ of heat is evolved. For this process, calculate
    (a) Q   (b) ΔE
    (c) W   (d) Δn_g
    (e) ΔH

*39. From the data at 298 K, 1 atm:

	ΔH (kJ)
$\frac{1}{2}H_2(g) + \frac{1}{2}O_2(g) \rightarrow OH(g)$	+42.09
$H_2(g) + \frac{1}{2}O_2(g) \rightarrow H_2O(g)$	−241.80
$H_2(g) \rightarrow 2\ H(g)$	+435.89

calculate ΔH for the reaction $H(g) + OH(g) \rightarrow H_2O(g)$.

*40. Consider the reaction, at room temperature and atmospheric pressure, $H_2(g) + Cl_2(g) \rightarrow 2\ HCl(g)$. In the presence of UV light, this reaction may proceed explosively, by means of a chain reaction. It is thought that the first step occurring at the level of individual molecules involves the splitting of a reactant molecule to form two highly reactive atoms. What do you suppose to be the most likely initial step? Why?

SELF-TEST ANSWERS

1. **F** (To restore the temperature to its initial value, heat must eventually be added to the test tube.)
2. **T** (Another way of looking at Question 1.)
3. **F** (Absorbed.)
4. **T** (Often seen in other texts.)
5. **F** (The first heat effect would be that for forming one mole of water; the second, twice as much heat for twice the amount of water. This is how the equation stoichiometry is interpreted.)
6. **T**
7. **F** (It is solid carbon that is undergoing combustion, but gaseous carbon dioxide that is being formed from the elements.)
8. **F** (The given reaction can be considered as the sum of the two reactions: $H_2(g) + \frac{1}{2} O_2(g) \rightarrow H_2O(l)$; $2 H(g) + O(g) \rightarrow H_2(g) + \frac{1}{2} O_2(g)$. The second step releases additional energy, making the overall ΔH more negative.)
9. **F** (Q could be practically anything, provided other kinds of energy transfer, such as work of expansion, were involved.)
10. **T** (You probably expect this from common experience: heat is generally evolved during reactions. For a closer look at spontaneous reactions, see Chapter 14.)
11. **c** (ΔH is positive for $H_2O(l) \rightarrow H_2O(g)$; additional energy is stored in the gas. So, steam causes the more severe burns.)
12. **c** (Evaporation requires heat; heat is removed from the hot body.)
13. **d** (Note that CO_2 is generally formed, and not CO, in the reaction chemists call combustion.)
14. **c**
15. **d**
16. **b** (Choice c, for example, is ruled out since ΔH doesn't change much with temperature.)
17. **b**
18. **a** (Check your units as well as the sign.)
19. **c** (The energy released increases from (a) to (b) to (c); as well as from (d) to (c), since burning CO to form CO_2 would release the larger amount of energy associated with (c). Do you also see the rationale for always specifying the states?)

56 • 4–Thermochemistry

20. b
21. a (First note that the given enthalpy changes are heats of *combustion*. ΔH = +1.9 kJ/mol. This is small compared to most heats of reaction for chemical changes. So why were diamonds only recently synthesized?)
22. c (See Readings.)
23. b (See Readings.)
24. d (I.e., fusion reactors or solar energy.)
25. d
26. d (To find: ΔH for the reaction HCl(g) → H(g) + Cl(g).)
27. b (The stronger bond is the more difficult to break; i.e., ϵ is larger.)
28. b (Energy is required to break any bond.)
29. d (And not (a), since the elements are not in their stable, diatomic form; not (b) or (c) because the product is liquid instead of gas.)
30. d (This is the understanding of bonding required for this chapter; more in Chapter 8.)
31. d (From another point of view: specific values of P, T, and composition will automatically determine the values of H, E, V, and other *state properties*.)
32. c (This is one reason for discussing the First Law.)
33. c (20 g × 4 J/(g·°C) × 10°C = +800 J.)
34. d (Final and initial states are one and the same. See answers for 31 and 9.)
35. b (By definition, ΔH = ΔE + PΔV, for constant P; 2.5 kJ = PΔV = work of expansion.)
36. Per mole C_3H_8: ΔH = 3(−393.5 kJ) + 4(−285.8 kJ) − (−103.8 kJ)
    = −2219.9 kJ

    1.40 g C_3H_8 × $\dfrac{-2219.9 \text{ kJ}}{44.0 \text{ g}}$ = −70.6 kJ

37. (a) −2599.9 kJ = 4(−393.5 kJ) + 2(−285.8 kJ) − 2ΔH_f C_2H_2(g)
    ΔH_f C_2H_2 = +227.2 kJ/mol

    (b) 2.60 × 10⁶ J = 5.00 × 10⁴ g × 4.18 $\dfrac{J}{g \cdot °C}$ × Δt; Δt = 12.4°C
    final t = 32.4°C

38. (a) −89.5 kJ (b) −89.5 kJ (c) 0 (d) +3 (e) −82.0 kJ

*39. H(g) → $\tfrac{1}{2}$ H_2(g)   ΔH₁ = −217.95 kJ

    OH(g) → $\tfrac{1}{2}$ H_2(g) + $\tfrac{1}{2}$ O_2(g)   ΔH₂ = −42.09 kJ

    H_2(g) + $\tfrac{1}{2}$ O_2(g) → H_2O(g)   ΔH₃ = −241.80 kJ

    ΔH = ΔH₁ + ΔH₂ + ΔH₃ = −501.84 kJ

*40. The more easily split molecule has the smaller bond energy: Cl_2(g) → 2 Cl(g). Thermodynamics provides clues for the mechanism by which a reaction proceeds. (See Chapter 16.)

SELECTED READINGS

Energy sources are considered in:

Bethe, H. A., The Necessity of Fission Power, *Scientific American* (January 1976), pp. 21–31.

Cheney, E. S., U.S. Energy Resources: Limits and Future Outlook, *American Scientist* (January–February 1974), pp. 14–22.

Daniels, F., *Direct Use of the Sun's Energy*, New Haven, Yale, 1964.

"Energy and Power," *Scientific American* (September 1971).

Holdren, J., *Energy*, New York, Sierra Club, 1971.

"Hydrogen: Likely Fuel of the Future," *Chemical and Engineering News* (June 26, 1972), pp. 14–17.

Poole, A. D., Flower Power: Prospects for Photosynthetic Energy, *Bulletin of the Atomic Scientists* (May 1976), pp. 49–58.

Walters, E. A., An Overview of the Energy Crisis, *Journal of Chemical Education* (May 1975), pp. 282–288.

Thermodynamics is introduced in three completely different ways in:

Faraday, M., *The Chemical History of a Candle*, New York, Viking, 1960.

Mahan, B. H., *Elementary Chemical Thermodynamics*, New York, W. A. Benjamin, 1963.

Pimentel, G. C., *Understanding Chemical Thermodynamics*, San Francisco, Holden-Day, 1969.

5

THE PHYSICAL BEHAVIOR OF GASES

QUESTIONS TO GUIDE YOUR STUDY

1. What materials can you think of that usually exist as gases? What properties do they have in common?

2. What conditions favor the existence of a substance as a gas rather than as a liquid or a solid?

3. How do you measure properties of gases such as temperature and pressure? How would you weigh a sample of gas?

4. Is there a simple relationship among the properties of a gas that generally holds for all gases? Can you rationalize such a relationship in terms of atomic-molecular theory? (For example: how does the behavior of molecules explain the relation between temperature and pressure for the air inside a tire?)

5. How do mixtures of gases behave? How is their behavior related to that of a pure gaseous substance?

6. Can the quantities of gases participating in a chemical reaction be simply expressed in terms of masses, moles, or volumes?

7. How does the volume of gaseous reactant or product depend on reaction conditions?

8. What are some of the practical applications (as well as support for other chemical principles) of our knowledge of gas behavior?

9. What experimental support is there for our ideas about the nature of the molecules in a gas?

10. How do you account for the observed differences in properties among gases and between gases and liquids and solids?

11.

12.

YOU WILL NEED TO KNOW

Concepts

1. Meaning of molecular mass and how to calculate it from a formula – Chapter 2.

Math

1. How to solve first and second order equations for any one variable. Examples: Rewrite the equation PV = gRT/(GMM) in the form GMM = ...

Solve the equation $\frac{1}{3}$(KMM)u² = RT for u.

2. How to find the square root of a number (most simply done by using a calculator, slide rule, or table of logarithms). Note that square roots are eliminated from an equation when both sides of the equation are squared. See Skill 4 below; also see Readings listed in the Preface.

CHAPTER SUMMARY

The physical behavior of gases is described concisely by the Ideal Gas Law, PV = nRT, which forms the central theme of this chapter. This equation tells us how the four experimental variables, pressure, volume, number of moles, and absolute (Kelvin) temperature, are related to one another. We see, for example, that for a given sample of gas at a fixed temperature (i.e., n and T constant), PV = constant, or P = constant/V. That is, the Ideal Gas Law embodies Boyle's Law. Again, we see that for a given sample of gas at a fixed pressure (i.e., P and n constant), V = constant X T (Law of Charles and Gay-Lussac). In still another case, we see that when temperature and pressure are held constant, V = constant X n, which implies Avogadro's Law (equal volumes of all gases at the same temperature and pressure contain equal numbers of molecules).

Frequently, we use the Ideal Gas Law to calculate one variable given the values of the others. In order to carry out such calculations, we must know the magnitude of the gas law constant, R. For our purposes in this chapter, it is most convenient to express R as:

$$R = 8.31 \text{ kPa} \cdot \text{dm}^3/(\text{mol} \cdot \text{K})$$

Perhaps the greatest single advantage of using SI units in general chemistry is that R retains the same numerical value regardless of the type of calculation in which it is involved. Thus, we have:

$$R = 8.31 \text{ J}/(\text{mol} \cdot \text{K}) = 8.31 \text{ kg} \cdot \text{m}^2/(\text{s}^2 \cdot \text{mol} \cdot \text{K})$$

For certain applications of the Ideal Gas Law, it is convenient to substitute for the number of moles, n, its equivalent in grams, i.e., n = g/GMM), where g is the number of grams of the gas and GMM is its gram molecular mass. The resulting equation, PV = gRT/(GMM), can be used to determine the molecular mass of a gas from measured values of P, V, g, and T. Furthermore, recognizing that the density is mass divided by volume, we can obtain a relation which tells us, among other things, that the densities of different gases at the same temperature and pressure are in the same ratio as their molecular masses.

The Ideal Gas Law can be applied to gas mixtures as well as to pure gases. We can, for example, write $P_A V = n_A RT$, where P_A is the partial pressure and n_A the number of moles of gas A in the mixture. By combining similar expressions for each gas in the mixture, it is possible to obtain Dalton's Law. This law is particularly useful in making calculations involving "wet" gases. Such mixtures, in which water vapor is one component, are commonly formed in the laboratory when gases are collected over water.

The validity of the Ideal Gas Law is confirmed by experimental measurements which require no assumptions about the behavior of gas molecules. However, by making some rather simple assumptions concerning molecular motion, embodied in what is known as kinetic molecular theory, it is possible to derive the law from "first principles." This exercise was one of the great triumphs of nineteenth century science: it provided the first really convincing evidence for the existence of molecules and atoms. A key postulate of kinetic theory is that, at a given temperature, molecules of all gases have the same kinetic energy of translation. Specifically, $\epsilon = mu^2/2 =$ constant $\times$ T, where m is the mass and u the average speed of a molecule. Starting with this postulate, it is possible to derive equations for the relative rates of effusion of different molecules (Graham's Law) or the average molecular speed of a particular gas at a given temperature. You should keep in mind that u in these equations is an *average* speed; at any given instant, virtually all of the molecules in a gas sample are moving at speeds either greater or smaller than u. This statistical range or distribution of speeds results, as we intuitively expect, from collisions between molecules: a collision between two molecules will change the velocities of the two particles, in terms of magnitude, generally, as well as direction.

We will see in later chapters (particularly 11 and 16) how important is the existence of the distribution of energies and speeds among molecules. The fact that some molecules have considerably greater than average energy will be important in our interpretation of such seemingly diverse phenomena as the evaporation of a liquid and the way in which reaction rate changes with temperature.

The simple kinetic molecular model of gases ignores interactions between molecules and assumes their volume to be negligible in comparison to that of their container. At low pressures and high temperatures, where the molecules are far apart, these approximations are justified and we find that the Ideal Gas Law describes very well the behavior of real gases. However, at high pressures and low temperatures, intermolecular forces and molecular volumes become significant and real gases deviate considerably from ideal behavior. Indeed, if the molecules of a gas approach each other closely enough, condensation to a liquid or solid occurs, and the Ideal Gas Law no longer applies.

BASIC SKILLS

You will find that the material in this chapter can best be mastered by working problems. In particular, you should be able to perform the following skills.

1. Use the Ideal Gas Law to:

    a. Determine the effect of a change in conditions (e.g., a change in T or P) upon a particular variable (e.g., V).

A sample of gas has a volume of 312 cm^3 at 273 K and 103 kPa. What volume will the same sample occupy at 298 K and 101 kPa?

For the gas in both states PV = nRT. But, from the statement of the problem, the number of moles of gas, n, remains constant. The gas constant, R, must of course remain unchanged. It follows that the quantity PV/T must have the same value in the final and initial states. That is,

$$\frac{P_2 V_2}{T_2} = \frac{P_1 V_1}{T_1}$$

where the subscript 1 refers to the initial state and 2 to the final state. Solving for the variable we need, V_2,

$$V_2 = \frac{P_1 V_1}{T_1}\left(\frac{T_2}{P_2}\right) \quad \text{or,} \quad V_2 = V_1 \times \frac{T_2}{T_1} \times \frac{P_1}{P_2}$$

Having derived the relationship required, all that remains is to substitute numbers. From the statement of the problem, $V_1 = 312$ cm^3, $T_2 = 298$ K, $T_1 = 273$ K, $P_1 = 103$ kPa. $P_2 = 101$ kPa.

$$V_2 = 312 \text{ cm}^3 \times \frac{298 \text{ K}}{273 \text{ K}} \times \frac{103 \text{ kPa}}{101 \text{ kPa}} = 347 \text{ cm}^3$$

This technique is further illustrated in Examples 5.1 and 5.2. Note that we can always derive the relationship required, which may involve two variables (e.g., P and V in Example 5.1, V and T in Example 5.2) or three variables (V, T, and P in the example above), from the Ideal Gas Law by applying simple algebra. In case you are tempted to try to memorize these relationships, we should point out that there are ten of them! Notice that in order to solve problems of this type, both T_2 and T_1 must be in K. Furthermore, P_2 and P_1 (or V_2 and V_1) must be expressed in the same units. We could, for example, express both P_2 and P_1 in atmospheres or both in kilopascals, but we could not use P_2 in atm and P_1 in kPa.

Problems 5.1a, b, 5.5, 5.7, 5.24, and 5.26 are entirely analogous to the examples cited; Problem 5.27 introduces a minor complication (gauge pressure).

      b. Solve for one variable (e.g., V) given the values of the other three (e.g., n, P, and T).

What volume is occupied by 2.10 mol of an ideal gas at 20°C and 150 kPa? _____
Solving the Ideal Gas Law for V:

$$V = \frac{nRT}{P}$$

Substituting n = 2.10 mol, R = 8.31 kPa·dm^3/(mol·K), T = (20 + 273) = 293 K, P = 150 kPa:

$$V = \frac{(2.10 \text{ mol})[8.31 \text{ kPa·dm}^3/(\text{mol·K})](293 \text{ K})}{150 \text{ kPa}} = 34.1 \text{ dm}^3$$

Example 5.3 is analogous except that you must first convert grams of XeF$_4$ to moles. Example 5.4 is slightly more complex, requiring the conversion of moles to kilograms in the final step. Of problems of this type

64 • 5–The Physical Behavior of Gases

at the end of Chapter 5, 5.1c, 5.9, 5.28, and 5.29 are straightforward; 5.10 is a little more difficult.

Note that in this application of the Ideal Gas Law, the units used for the variables must be consistent with the value chosen for R. It is certainly simplest to stay with a single value of R as 8.31 kPa·dm^3/(mol·K) and, if necessary, convert P to kilopascals or V to cubic decimetres.

This skill can be combined with the principles regarding mass relations in chemical reactions (Chapter 3) to relate the number of grams or moles of one species to the volume of a gaseous reactant or product. This application is shown in Example 5.7 and is illustrated in Problems 5.15 and 5.34. Problems 5.16 and 5.35 follow the same principle but are somewhat more difficult; in both cases, a reasonable first step is to calculate the number of moles of hydrogen.

c. Calculate the density of a gas at a given temperature and pressure.

Here, and for (d) below, it is convenient to write the Ideal Gas Law in the form

$$PV = gRT/(GMM) \qquad \text{(Equation 5.8)}$$

Realizing that density is the ratio of mass (g) to volume (V), we obtain the general relation

$$\text{density} = \frac{g}{V} = \frac{(GMM)P}{RT}$$

The use of this relation is illustrated in Example 5.5 and Problems 5.1d and 5.30. Problem 5.11 is a little more subtle in that you first have to calculate an effective molecular mass for air.

d. Calculate the molecular mass of a gas, knowing the mass of a given volume (or the density) at a known P and T.

See Example 5.6 and Problems 5.12 and 5.31.

2. Relate the volumes of gases (measured at the same P and T) in a chemical reaction.

The basic principle here is a simple one: the coefficients of a balanced equation relate not only the numbers of moles of reactants and products, but also the volumes of gases measured at the same temperature and pressure. Thus for the reaction

$$2 H_2(g) + O_2(g) \rightarrow 2 H_2O(g)$$

if we started with 50 dm³ of H_2, 25 dm³ of O_2 would be required to react with it and 50 dm³ of water vapor would be produced, all at the same temperature and pressure. This principle is applied directly to solve Problems 5.14 and 5.33.

A note of caution: this simple relationship between reacting volumes is restricted to *gaseous* reactants or products at the *same temperature and pressure*. In the example just cited, if liquid water had been involved, its volume would have been a great deal less than 50 dm³. Again, if the gases were at different temperatures or pressures, the 2:1:2 volume relationship would not hold.

3. **Use Dalton's Law to obtain partial pressures of gases in mixtures.**

This skill is shown in Example 5.8 and applied in Problem 5.2, which is entirely analogous. Problems 5.17 and 5.36 are basically similar but somewhat more difficult. Quite frequently, Dalton's Law is used to obtain the partial pressure of a gas collected over water, as in the following example.

A sample of H_2 is collected over water at a total pressure of 98.2 kPa. The partial pressure of water vapor in the gaseous mixture is 3.2 kPa. What is the partial pressure of H_2? _____
Applying Dalton's Law:

$$P_{total} = P_{H_2} + P_{H_2O}$$

$$P_{H_2} = P_{total} - P_{H_2O} = 98.2 \text{ kPa} - 3.2 \text{ kPa} = 95.0 \text{ kPa}$$

This simple operation is often the first step in a gas law calculation. In Example 5.9, we first obtain the partial pressure of H_2 and then use the Ideal Gas Law to obtain the number of moles of H_2. Problems 5.13 and 5.32 are similar.

4. **Use Graham's Law to relate the molecular masses of two gases to their rates of effusion.**

It is found that the rate of effusion of a certain gas is 0.600 times that of O_2 under the same conditions. What is the molecular mass of the gas?

_____

Applying Graham's Law

$$\frac{r_X}{r_{O_2}} = \left(\frac{MM_{O_2}}{MM_X}\right)^{1/2}$$

Now, since $r_X = 0.600\ r_{O_2}$, we see that the ratio r_X/r_{O_2} is 0.600. The molecular mass of O_2 is 32.0. Hence,

$$0.600 = \left(\frac{32.0}{MM_X}\right)^{1/2}$$

Squaring both sides and solving for MM_X:

$$0.360 = \frac{32.0}{MM_X}\ ;\ MM_X = \frac{32.0}{0.360} = 88.9$$

In many problems of this type, the times of effusion rather than the rates are specified (Example 5.11; Problems 5.21 and 5.40). The time required for an event to take place is inversely related to the rate at which it occurs.

5. Calculate the average speed of molecules of a particular gas at a given temperature.

This calculation is illustrated in Example 5.10 and in Problem 5.39. Note that here we must use the "kilogram molecular mass," KMM. For oxygen, for example, KMM = 0.0320.

SELF-TEST

True or False

1. To prepare pure nitrogen gas from a sample of air, one ( ) would probably employ fractional distillation.

2. The molecules of a sample of any gas, under most ( ) conditions, can be characterized as being separated by relatively large distances.

3. At a temperature of 20°C, just about 3.4 g of carbon ( ) dioxide gas will dissolve in 1.00 dm³ of water under a pressure of one atmosphere. At a higher temperature, but at the same pressure, one would expect that more carbon dioxide will dissolve.

4. The pressure exerted by one mole of O_2 in a 10.0 dm³   (   )
container at 27°C will be 1.00(8.31)(300)/10.0 kPa.

5. In a mixture of two gases, with a total pressure of 760 kPa,   (   )
the partial pressure of chlorine gas is 380 kPa. This means that half
of the molecules in the sample are chlorine.

6. The average kinetic energy of an oxygen molecule and the   (   )
average kinetic energy of a hydrogen molecule, both gases at the
same temperature, are in the same ratio as their molecular masses.

7. The average speed of a gas molecule depends only on the   (   )
absolute temperature.

8. Two separate samples of the same gaseous substance at the   (   )
same pressure would have densities in the same ratio as their absolute
temperatures.

9. The fact that a sample of gas would not have zero volume at   (   )
the absolute zero of temperature is a consequence of the fact that
absolute zero cannot be reached.

10. In the process known as effusion, two or more gaseous   (   )
substances mix to give a solution.

Multiple Choice

11. The chemical analysis of a mixture of gases is most likely to   (   )
involve
    (a) density measurements
    (b) mass spectrometry
    (c) boiling point measurements
    (d) fractional crystallization

12. Which one of the following substances would you expect to   (   )
normally exist as a gas at room temperature?
    (a) CH_4 (b) C_5H_{12}
    (c) C_6H_6 (d) C_3H_7OH

13. The volume of a mole of gas is   (   )
    (a) 22.4 dm³
    (b) directly proportional to pressure and absolute temperature
    (c) directly proportional to pressure, inversely proportional to absolute temperature
    (d) inversely proportional to pressure, directly proportional to absolute temperature

68 • 5-The Physical Behavior of Gases

14. The inflation of an automobile tire to a pressure of "20 ( )
pounds" means that
    (a) the mass of the air in the tire is 20 pounds
    (b) the pressure exerted by the air in the tire is 20 lb/in^2
    (c) the pressure of the air in the tire is 20 lb/in^2 higher than the air pressure outside the tire
    (d) the pressure of the air inside the tire is about 5 lb/in^2

15. A certain mountain rises to 5000 m above sea level. The ( )
pressure at the top is about 17.7 inches (of mercury). If you blew up a balloon at sea level, where the pressure happened to be 29.7 inches, and carried it to the top of the mountain, by what factor would its volume change?
    (a) there would be no change (b) 29.7 - 17.7
    (c) 29.7/17.7   (d) 17.7/29.7

16. When equal numbers of moles of two gases at the same ( )
temperature are mixed in a container, the pressure of the gaseous mixture is
    (a) given by Gay-Lussac's Law of combining volumes
    (b) the product of the pressures each gas would have if alone
    (c) the difference of the pressures each gas would have if alone in the container
    (d) the sum of the pressures each gas would have if alone

17. A gaseous substance is known which can be decomposed to ( )
give only the elements phosphorus and hydrogen. When all three substances are gases at a convenient temperature and pressure, it is found that four volumes of the compound give one volume of phosphorus and six volumes of hydrogen. The simplest interpretation is that phosphorus gas is
    (a) P   (b) P$_2$
    (c) P$_3$ (d) P$_4$

18. An equation representing a reaction which is consistent with ( )
the data of Question 17 would be:
    (a) 4 PH$_2$(g) → 2 P$_2$(g) + 4 H$_2$(g)
    (b) 4 PH$_3$(g) → P$_4$(g) + 6 H$_2$(g)
    (c) 2 PH$_3$(g) → 2 P(g) + 3 H$_2$(g)
    (d) some other equation

19. Two flasks of equal volume are filled with different gases, A ( )
and B, at the same temperature and pressure. The mass of gas A is 0.34 g while that of gas B is 0.48 g. It is known that gas B is ozone, O$_3$ and that gas A is one of the following. Which one is most likely

to be gas A?
- (a) O_2
- (b) H_2S
- (c) SO_2
- (d) cannot say

20. Which one of the following statements about the gases A ( ) and B of Question 19 is true?
    - (a) the numbers of molecules of A and B are equal
    - (b) the masses of individual molecules of A and B compare in the same way as the masses of the samples
    - (c) the average translational energies of molecules of A and B are the same
    - (d) all the above are true

21. The fact that the Ideal Gas Law only approximately ( ) describes the behavior of a gas can be partly explained by the idea that
    - (a) R is not really a constant
    - (b) gas molecules really do have zero volume
    - (c) the kinetic energy of gas molecules is not really directly proportional to the absolute temperature
    - (d) gas molecules really do interact with each other

22. Real gases behave most nearly like the Ideal Gas Law says ( ) they do at
    - (a) high temperatures, low pressures
    - (b) low temperatures, high pressures
    - (c) high temperatures, high pressures
    - (d) low temperatures, low pressures

23. The van der Waals equation, $P = RT/(V - b) - a/V^2$, ( ) incorporates the following correction(s) to the Ideal Gas Law in order to account for the properties of real gases:
    - (a) the possibility of chemical reaction between molecules
    - (b) the finite volume of molecules
    - (c) the quantum behavior of molecules
    - (d) average kinetic energy is inversely proportional to temperature

24. To convert a sample of air into a liquid, you would ( ) probably have to
    - (a) increase the temperature and pressure of the sample
    - (b) decrease the temperature and increase the pressure of the sample
    - (c) cool it to 0 K
    - (d) the task is an impossible one

5–The Physical Behavior of Gases

25. Your lecturer opens a bottle of hydrogen sulfide gas, H_2S, ( ) and a bottle of gaseous diethyl ether, $C_4H_{10}O$, at the same time. Both gases are at the same temperature and pressure. Which of the two should you be able to smell first? (Both have characteristic odors.)
    (a) the ether
    (b) H_2S
    (c) both at the same time
    (d) neither gas would escape into the room, since they are both heavier than air

26. What must be the molecular mass of a gas that effuses ( ) one-fourth as rapidly as CH_4 (MM = 16.0)?
    (a) 4
    (b) 16
    (c) 64
    (d) 256

27. When 6.00 dm³ of N_2 and 6.00 dm³ of H_2 are mixed, the ( ) following reaction occurs: $N_2(g) + 3\ H_2(g) \rightarrow 2\ NH_3(g)$. What volume of NH_3 is produced at the same temperature and pressure at which these volumes of reactants were measured? (Assume 100% yield.)
    (a) 2.00 dm³
    (b) 4.00 dm³
    (c) 6.00 dm³
    (d) 12.0 dm³

28. Which molecular property tends to produce large deviations ( ) from ideal gas behavior?
    (a) high molecular speed
    (b) small molecular mass
    (c) large molecular volume
    (d) weak intermolecular attraction

29. In a mixture of H_2, He, and Ne at 25°C, the molecules ( ) moving with the greatest average speed are
    (a) H_2
    (b) He
    (c) Ne
    (d) all speeds will be the same

30. At room temperature, the translational energy of a mole of ( ) gas is about
    (a) 1 J
    (b) 2 J
    (c) 4 kJ
    (d) 8×10^7 kJ

Problems

31. A CO_2 fire extinguisher having a volume of 5.67 dm³ weighs 9.43 kg when full and 6.93 kg when empty. Assuming CO_2 (MM = 44.0) behaves as an ideal gas, calculate the pressure of CO_2, in kilopascals, in the fire extinguisher when full at 33°C.

32. A 5.00-dm³ flask contains a mixture of NO (MM = 30.0) and C_3H_6 (MM = 42.0) at 27°C and a total pressure of 150 kPa. Analysis shows 2.10 g of C_3H_6 to be present.
   (a) Determine the partial pressure of C_3H_6.
   (b) Determine the partial pressure of NO.

33. A volatile liquid is completely vaporized at 102°C in a 315-cm³ flask. Barometric pressure is found to be 98.9 kPa. Upon cooling the flask, it is found that the vapor weighs 0.887 g. What is the molecular mass of the liquid?

*34. A 0.680-g sample of phosphine, PH_3 (MM = 34.0), occupies a volume of 589 cm³ at 77.0°C and 98.9 kPa. When completely decomposed to the elements at this temperature and pressure, the measured volume is 1030 cm³.
   (a) Calculate the total number of moles of gaseous products.
   (b) *Based on the information given* and the fact that gaseous hydrogen is H_2, write the balanced equation for the reaction. Be sure that the equation does indeed fit the data!

*35. In the text discussion of the distribution of molecular speeds, it is noted that, in $O_2(g)$ at 25°C, only about one fifth as many molecules have a speed of 800 m/s as have a speed of 400 m/s. Show this to be so by calculation.

The relative number of molecules with speed u is given by the expression

$$n_u = 4\pi u^2 \left(\frac{KMM}{2\pi RT}\right)^{3/2} e^{-(KMM)u^2/2RT}$$

SELF-TEST ANSWERS

1. **T** (The air would be liquefied and then fractionally distilled. See Chapters 1 and 17.)
2. **T** (Large compared to molecular diameters.)
3. **F** (Think of what would happen if you warmed up an opened bottle of carbonated water, or of Coke. However, not all gas-liquid solutions behave this way — see Chapter 12.)
4. **T** (P = nRT/V)
5. **T**
6. **F** (They would be the same.)
7. **F** (Also on molecular mass.)
8. **F** (Inversely, density = mass/volume = P(GMM)/RT. So, D(A)/D(B) = T(B)/T(A) for gases A and B.)

9. F (Any real gas condenses to a liquid or solid well above absolute zero.)
10. F (Effusion: a gas leaks into a vacuum.)
11. b
12. a (Low molecular mass – Chapter 9.)
13. d (The volume would be 22.4 dm³ only under standard conditions, 0°C and 1 atm.)
14. c (Now you know!)
15. c (Larger volume at lower pressure.)
16. d (Dalton's Law of Partial Pressures.)
17. d (See the next question, also based on the laws of Avogadro and Gay-Lussac.)
18. b
19. b (Molecular masses must also compare 0.34/0.48, so MM of A is 34.)
20. d
21. d (Attractive forces are particularly important as the temperature approaches the bp.)
22. a (High T means compared to the bp; low P usually means $\leq$ 1 atm.)
23. b (In the term V – b; intermolecular attractions, in the term a/V^2.)
24. b (These are the conditions under which the real gas deviates most from the gas laws.)
25. b (With the lower MM, and given the same average translational energies, the H_2S would have the higher average speed. Given enough time, both would fill the room.)
26. d (Substitute into Equation 5.17.)
27. b (H_2 is the limiting reagent – Chapter 3.)
28. c
29. a (With the smallest mass, H_2 molecules must move faster to have the same energy as either He or Ne.)
30. c
31. $P = \dfrac{nRT}{V}$; $n = \dfrac{2.50 \times 10^3 \text{ g}}{44.0 \text{ g/mol}} = 56.8$ mol

$P = \dfrac{(56.8 \text{ mol})[8.31 \text{ kPa} \cdot \text{dm}^3/(\text{mol} \cdot \text{K})](306 \text{ K})}{5.67 \text{ dm}^3} = 25\,500$ kPa

32. (a) $n\,(C_3H_6) = \dfrac{2.10 \text{ g}}{42.0 \text{ g/mol}} = 0.0500$ mol

$P\,(C_3H_6) = \dfrac{(0.0500 \text{ mol})[8.31 \text{ kPa} \cdot \text{dm}^3/(\text{mol} \cdot \text{K})](300 \text{ K})}{5.00 \text{ dm}^3}$

$= 24.9$ kPa

(b) P (NO) = 150 kPa – 25 kPa = 125 kPa

33. $\text{GMM} = \dfrac{gRT}{PV} = \dfrac{(0.887 \text{ g})[8.31 \text{ kPa·dm}^3/(\text{mol·K})](375 \text{ K})}{(98.9 \text{ kPa})(0.315 \text{ dm}^3)}$

$= 88.7 \dfrac{\text{g}}{\text{mol}}$

MM = 88.7

34. This is a restatement of Questions 17 and 18.

    (a) $n \text{ (products)} = \dfrac{PV}{RT} = 0.0350; \; n \text{ (PH}_3) = \dfrac{0.680}{34.0} = 0.0200$

    (b) 1 mol $PH_3 \to$ 1.75 mol products;

    4 mol $PH_3 \to$ 7 mol products; $4 PH_3(g) \to P_4(g) + 6 H_2(g)$

35. Dividing the number of molecules at 800 m/s by the number at 400 m/s and cancelling $4\pi \left(\dfrac{\text{KMM}}{2\pi RT}\right)^{3/2}$

$\dfrac{n_{800}}{n_{400}} = \dfrac{(8 \times 10^2)^2 \exp[-(\text{KMM})(8 \times 10^2)^2/2RT]}{(4 \times 10^2)^2 \exp[-(\text{KMM})(4 \times 10^2)^2/2RT]}$

$= 4 \exp\left[\dfrac{\text{KMM}}{2RT}(16 \times 10^4 - 64 \times 10^4)\right] = 4e^{-3.10}$

$= 4(0.045)$

$= 0.18$

SELECTED READINGS

History of investigations into gas behavior is considered in:

Conant, J. B., *Science and Common Sense*, New Haven, Yale, 1951.
Neville, R. G., The Discovery of Boyle's Law, 1661–62, *Journal of Chemical Education* (July 1972), pp. 356–359.

Kinetic theory is extensively discussed, with lots on real behavior, in:

Hildebrand, J. H., *An Introduction to Modern Kinetic Theory*, New York, Reinhold, 1963.
Kauzmann, W., *Kinetic Theory of Gases*, Menlo Park, Ca., W. A. Benjamin, 1966.

Tools for studying gases at low pressures:

Steinherz, H. A., Ultrahigh Vacuum, *Scientific American* (March 1962), pp. 78–90.

6
THE ELECTRONIC STRUCTURE OF ATOMS

QUESTIONS TO GUIDE YOUR STUDY

1. What evidence is there for the idea that atoms are themselves composed of smaller parts? What are the dimensions of a typical atom and its parts? (Review Chapter 2.)

2. What differences are there in the behavior of electrons inside atoms versus outside? (How do electrons composing cathode rays or β-particles compare to atomic electrons?)

3. For an electron bound to (inside) an atom—where is it, what is it doing?

4. How many arrangements are possible for the electron(s) in a hydrogen atom? In a copper atom? Which are observed under "ordinary" conditions?

5. How are electron arrangements in atoms represented?

6. What experimental observations support the details of electron arrangement theory?

7. What energy changes are associated with changes in electron structure? How much energy is required to remove an electron from a neutral atom?

8. How do electron and thermochemical energies compare?

9. What determines the number of electrons a neutral atom will possess? The number of electrons the atom may lose or gain to form an ion?

10. Does modern quantum theory provide all the answers?

11.

12.

YOU WILL NEED TO KNOW

Concepts

1. General ideas of atomic and nuclear composition (kinds, numbers, and charges of nuclear particles; meaning of atomic number)—Review Chapter 2.
2. A general idea of what composes the electromagnetic spectrum (infrared, visible, ultraviolet, x-rays . . .)—the names of the regions and their approximate energies, wavelengths, and frequencies—See this chapter, as well as any introductory physics textbook; also see Table 1 in Chapter 8 of this guide.

Math

1. No new math is introduced here.
2. How to relate the charge of a monatomic ion, the number of electrons, and the nuclear charge (or atomic number)—Chapter 2.

CHAPTER SUMMARY

The electronic structures of atoms, the major theme of this chapter, have occupied the attention of chemists and physicists for more than a century. As so often happens, theories have been developed and subsequently modified to explain puzzling experimental observations. In particular, it was found that light emitted by "excited" gaseous atoms possesses certain discrete wavelengths. To explain this phenomenon, it was proposed that electrons in atoms are restricted to discrete energy levels whose separation, ΔE, is related to the wavelength, λ, of the emitted light by the Einstein equation: $\Delta E = hc/\lambda$ (for λ in metres and ΔE in joules per atom, $hc = 1.99 \times 10^{-25}$). This postulate of the quantization of electronic energies is basic to modern quantum theory.

In 1913, Niels Bohr derived an equation for the possible energy levels in a one-electron atom. This equation agreed remarkably well with the observed spectrum of hydrogen. The Bohr model assumed that the electron, in moving about the nucleus, would be restricted to discrete circular orbits of fixed radius. Unfortunately, this simple model breaks down for any species with more than one electron. Modern quantum theory tells us that exact positions of electrons cannot be specified. The best we can hope to do is to calculate the probability of finding an electron in a particular region.

Electronic energies can be calculated, at least in principle, from the Schrödinger wave equation. You were probably relieved to find that we did not attempt to solve this formidable mathematical expression for allowed energies. Instead, we discussed briefly a simpler, less general expression, Equation 6.10, corresponding to the so-called "particle in a box" model. This model helps us to understand why small particles confined to very small regions of space do not obey the classical laws of motion with which we are familiar.

To completely characterize an electron in an atom, we indicate its:

1. *principal energy level*, specified by the quantum number **n**, which is restricted to positive, integral values (**n** = 1, 2, 3, . . .);

2. *sublevel*, designated by the letters s, p, d, f, ... Within a principal level of quantum number **n**, there are n sublevels. For **n** = 1, we have only the s sublevel; the second principal energy level has two sublevels (2s, 2p); the third, three (3s, 3p, 3d); and so on. This same information may be expressed by assigning each sublevel a quantum number ℓ and stating that ℓ can take on any integral value from 0 to (n − 1);

3. *orbital.* Within a sublevel of quantum number ℓ, there are $2\ell + 1$ orbitals. An s sublevel ($\ell = 0$) has only one orbital, a p sublevel ($\ell = 1$) has three, a d sublevel ($\ell = 2$) five, and a f sublevel ($\ell = 3$) seven. Orbitals are assigned quantum numbers, **m**$_\ell$, which take on all integral values from $-\ell$ to ℓ;

4. *spin*, which is restricted to two possible values. These may be indicated as "up" and "down" (↑↓) or by assigning a quantum number **m**$_s$ of +1/2 or −1/2. Experimentally, we find that no two electrons in an atom can have the same set of four quantum numbers (Pauli exclusion principle); it follows (why?) that a given orbital can contain no more than two electrons.

For many purposes, it is sufficient to describe the electronic structure of an atom by quoting its *electron configuration,* which tells us how many electrons are located in each sublevel. Thus, the electron configuration for nitrogen (at. no. = 7), $1s^2 2s^2 2p^3$, tells us that there are two electrons in the 1s sublevel, two in the 2s, and three in the 2p. Electron configurations can

78 • 6—The Electronic Structure of Atoms

be deduced by knowing the order in which sublevels are filled and recalling the total capacity of each sublevel (s = 2 × 1 = 2; p = 2 × 3 = 6; d = 2 × 5 = 10; f = 2 × 7 = 14). Examination of Table 6.4 shows us that this procedure is not infallible. Electrons occasionally show up in unexpected places (look at Cr and Cu, for example), but this need not concern us now.

Sometimes we need to go one step further and give the *orbital diagram* of an atom, which indicates the number of electrons in each orbital and their relative spins. To derive orbital diagrams from electron configurations, we need only take account of two factors: (1) two electrons in the same orbital have opposed spins; (2) whenever possible, orbitals within the same sublevel will be half-filled with electrons, all of the same spin (Hund's rule). Thus, the orbital diagram for nitrogen is

$$\begin{array}{ccc} 1s & 2s & 2p \\ (\uparrow\downarrow) & (\uparrow\downarrow) & (\uparrow)(\uparrow)(\uparrow) \end{array}$$

There are many questions we have yet to consider in discussing our modern model of atomic structure. Among them are: Why do ions form? (We need to interpret ionization potentials further.) Why do atoms bond together? Much of what composes the remainder of the text is an extension of atomic-molecular theory.

BASIC SKILLS

1. Use the Einstein equation (Equation 6.2) to relate the wavelengths of a spectral line to the difference in energy between two levels.

A typical calculation of this type is shown in Example 6.1; see also Problems 6.1, 6.7, and 6.23. Note that energies are sometimes required (or given) in units other than joules per particle. The following conversion factor will be helpful:

$$1 \text{ J/particle} = 6.02 \times 10^{20} \text{ kJ/mol}$$

2. Use the Bohr theory (Equation 6.5) to calculate the energy of an electron in a given principal energy level of the hydrogen atom, or the difference in energy between two levels.

See Example 6.3 and Problems 6.2, 6.9, and 6.25. Note that in Problem 6.2 the conversion factor listed above must be used; Problem 6.9 combines Skills 1 and 2. In using the Bohr equation, you should find that ΔE is positive when an electron moves from a low level (e.g., n = 1) to a higher level (e.g., n = 2) and negative when it moves in the reverse direction.

3. **Use the De Broglie relation (Equation 6.10) to obtain the minimum energy of a "particle in a box."**

Two typical calculations are given in the text (6.11 and 6.12). This skill is required in Problems 6.11, 6.12, 6.27, and 6.28. Note that in using Equation 6.10, the minimum energy will be obtained in joules when you substitute:

n = 1, h = 6.626 × 10⁻³⁴

m in kilograms (i.e., the mass of the particle must be expressed in kg)

d in metres

4. **Given the atomic number of an element, write the electron configuration of its atoms.**

The electron configuration of an atom gives the number of electrons in each sublevel. In order to write an electron configuration you must know:

a. the capacity of each sublevel (s = 2, p = 6, d = 10, f = 14).
b. the order in which various sublevels are filled in atoms (1s, 2s, 2p, 3s, 3p, 4s, 3d, . . .). The complete list given in p. *139* of the text need not be memorized; it can always be deduced from the position of the element in the Periodic Table (Chapter 7).

This skill is illustrated in Example 6.5; Problems 6.3 and 6.15 are entirely analogous. Problem 6.31 is similar but you first have to determine the number of electrons in K^+ and F^-

5. **Given, or having derived, the electron configuration of an atom, draw its orbital diagram.**

To make this conversion, you must realize that:

a. each sublevel is divided into orbitals capable of holding two electrons apiece. An s sublevel has one orbital, a p sublevel three orbitals, a d sublevel five, and an f sublevel seven.
b. When there are two electrons in an orbital, they have opposed spins, indicated by (↑↓).
c. In a partially filled sublevel, there are as many half-filled orbitals as possible. Electrons in these orbitals have the same spins, e.g., (↑) (↑) (↑).

Example 6.6 shows how the orbital diagrams of S and Ni atoms are derived from the electron configurations found in Example 6.5. See also

80 • 6–The Electronic Structure of Atoms

Problems 6.4, 6.17, and 6.33. In Problems 6.18 and 6.34 you are asked to go in the reverse direction, converting from orbital diagrams to electron configurations.

6. **Give the four quantum numbers corresponding to various electrons in an atom.**

The rules for assigning quantum numbers may be summarized as follows:

a. n = number of principal energy level = 1, 2, 3, 4, . . .
b. ℓ = 0, 1, . . . (n − 1). For an s electron, ℓ = 0; for a p electron, ℓ = 1; for a d electron, ℓ = 2; for an f electron, ℓ = 3.
c. m_ℓ can take on any integral value, including zero, ranging from $+\ell$ to $-\ell$.
d. m_s can be either +1/2 or −1/2.

This skill is illustrated in Example 6.7; Problems 6.19 and 6.35 give you a chance to apply it. Note that when a sublevel is partially filled, as is true in each of these cases, you have a choice of values for m_ℓ and m_s. For example, if there are five electrons in a p sublevel, the unpaired electron could be in an m_ℓ = 1, 0, or −1 orbital and could have a spin of +1/2 or −1/2. Thus, any of the following combinations would be correct for the p electrons in fluorine:

First	Second	Third	Fourth	Fifth
2, 1, 1, +1/2	2, 1, 1, −1/2	2, 1, 0, +1/2	2, 1, 0, −1/2	2, 1, −1, +1/2
2, 1, 1, +1/2	2, 1, 1, −1/2	2, 1, 0, +1/2	2, 1, 0, −1/2	2, 1, −1, −1/2
2, 1, 1, +1/2	2, 1, 1, −1/2	2, 1, −1, +1/2	2, 1, −1, −1/2	2, 1, 0, +1/2
2, 1, 1, +1/2	2, 1, 1, −1/2	2, 1, −1, +1/2	2, 1, −1, −1/2	2, 1, 0, −1/2
2, 1, 0, +1/2	2, 1, 0, −1/2	2, 1, −1, +1/2	2, 1, −1, −1/2	2, 1, 1, +1/2
2, 1, 0, +1/2	2, 1, 0, −1/2	2, 1, −1, +1/2	2, 1, −1, −1/2	2, 1, 1, −1/2

SELF-TEST

True or False

1. The volume of an atom is essentially that volume occupied ( ) by the electrons.

2. The greater the difference in energy between two levels, the ( ) longer the wavelength of the light emitted when an electron moves between them.

3. The atomic number is always equal to the number of ( ) electrons in a particular atom.

Self-Test • 81

4. The radius of a negatively charged monatomic ion is larger than the radius of the parent neutral atom. ( )

5. The energy associated with an electron in a given atom is almost fully described by specifying its value of the quantum number n. ( )

6. An orbital diagram is a geometrical representation of the shape of an orbital. ( )

7. The energy associated with electromagnetic radiation decreases in the order x-ray > ultraviolet > visible > infrared. ( )

8. When an atom absorbs a photon of energy E, the atom undergoes an increase in energy equal to or less than E. ( )

9. *Ground state* describes an atom with all its electrons in the lowest possible energy levels (and sublevels). ( )

10. Gravitational attraction makes a negligibly small contribution to the force holding an electron in an atom. ( )

Multiple Choice

11. What kind of attractive force holds together the components of an individual atom? ( )
    (a) gravitational   (b) magnetic
    (c) electrical   (d) the chemical bond

12. Experimental support for the arrangement of electrons in distinct energy levels is based primarily upon ( )
    (a) the law of constant composition
    (b) the law of conservation of energy
    (c) continuous spectra
    (d) spectra from electrical discharge through gases

13. For an electron with quantum number $\ell = 2$, the quantum number m_ℓ can have ( )
    (a) only one value
    (b) any one of three values
    (c) any one of five values
    (d) an infinite number of values

14. The number of electrons which can be accommodated in an electronic sublevel with $\ell = 2$ is ( )
    (a) 2   (b) 6
    (c) 10   (d) 14

15. What is the total electron capacity of the energy level for which n = 4? ( )

82 • 6–The Electronic Structure of Atoms

    (a) 8   (b) 16
    (c) 18   (d) 32

16. The possible values of the magnetic quantum number m_ℓ of ( )
a 3p electron are
    (a) 0, 1, 2   (b) 1, 2, 3
    (c) 1, 0, −1   (d) 2, 1, 0, −1, −2

17. The element whose neutral, isolated atoms have three ( )
half-filled 2p orbitals is
    (a) $_5$B   (b) $_6$C
    (c) $_7$N   (d) $_8$O

18. Which one of the following species has the same electron ( )
configuration as (is isoelectronic with) the argon atom?
    (a) Ne   (b) Na$^+$
    (c) S$^-$ (d) Cl$^-$

19. The electron configuration of the oxide ion, O^{2-}, may be ( )
represented as
    (a) $1s^2 2s^2 2p^4$ (b) $1s^2 2s^2 2p^2 3s^2 3p^2$
    (c) $1s^2 2s^2 2p^6$ (d) :O:$^{2-}$

20. Consider these four orbitals in a neutral calcium atom: 2p, ( )
3p, 3d, and the 4s. These orbitals arranged in order of increasing
energy are:
    (a) 2p < 3p < 3d < 4s   (b) 2p < 3p < 4s < 3d
    (c) 2p < 4s < 3p < 3d   (d) 4s < 2p < 3p < 3d

21. The electron structure of $^{40}_{20}$Ca^{2+} may be represented as ( )
    (a) $1s^2 2s^2 2p^6 3s^2 3p^6 4s^2$ (b) $1s^2 2s^2 2p^6 3s^2 3p^6$
    (c) $1s^2 2s^2 2p^6 3s^2 3p^6 3d^2$ (d) $1s^2 1p^6 1d^{10}$

22. Which one of the following species would have an odd ( )
number of electrons?
    (a) N
    (b) N$^+$
    (c) NO$_2^+$
    (d) electrons are always paired

23. The electron configuration of the Fe atom is [Ar] $4s^2 3d^6$. ( )
The number of unpaired electrons in the orbital diagram is
    (a) 0   (b) 2
    (c) 4   (d) 6

24. Which of the following represents a reasonable set of ( )
quantum numbers for a 3d electron?
    (a) 3, 2, 1, 1/2   (b) 3, 2, 0, −1/2
    (c) neither of these   (d) both of these

25. A proper set of four quantum numbers for the valence ( ) (outer-most) electron of rubidium, $_{37}$Rb, is
   (a) 5, 0, 0, +1/2
   (b) 5, 1, 0, +1/2
   (c) 5, 1, 1, +1/2
   (d) 6, 0, 0, +1/2

26. The most stable electron arrangement for the outer ( ) electrons (n = 2) of the $_5$B atom is
   (a) (↑↓) (↑) ( ) ( )
   (b) (↑) (↑) (↑) ( )
   (c) ( ) (↑) (↑) (↑)
   (d) ( ) (↑↓) ( ) (↑)

27. Which of the following could not be an orbital diagram for ( ) an atom in its ground state?
   (a) (↑↓) (↑↓) (↑↓) (↑) (↑)
       1s   2s   2p
   (b) (↑↓) (↑↓) (↑↓) (↑↓) (↑↓) (↑)
       1s   2s   2p   3s
   (c) (↑↓) (↑↓) (↑↓)(↑↓)( )
   (d) (↑↓) (↑↓) (↑↓)(↑↓)(↑↓) (↑↓)

28. Consider the successive ionization energies, E, of a lithium ( ) atom, $_3$Li:

$$E(1) + Li(g) \rightarrow Li^+(g) + e$$
$$E(2) + Li^+(g) \rightarrow Li^{2+}(g) + e$$
$$E(3) + Li^{2+}(g) \rightarrow Li^{3+}(g) + e$$

How should these ionization energies compare to each other?
   (a) E(1) = E(2) = E(3)
   (b) E(1) should be largest
   (c) E(2) should be largest
   (d) E(3) > E(2) ≫ E(1)

29. The removal of one electron from a neutral atom requires ( ) about the same energy for Li, Be, and B. The removal of a second electron, however, is expected to be much more difficult for
   (a) Li
   (b) Be
   (c) B
   (d) no more for one than the others

30. Consider the schematic energy level diagram shown for the hydrogen atom. Which one of the following interpretations is not correct? ( )

$n = \infty$ $E_\infty = 0$
$n = 3$ $E_3 = -146$ kJ/mol
$n = 2$ $E_2 = -328$ kJ/mol
$n = 1$ $E_1 = -1312$ kJ/mol

84 • 6—The Electronic Structure of Atoms

(a) The ground state and the first excited state differ in energy by (1312 - 328) kJ/mol.
(b) An electron in level E_1 is closer to the nucleus, on the average, than an electron in level E_2.
(c) Ionization is expected to occur with the absorption of at least 1312 kJ/mol.
(d) If an electron drops from level E_3 to level E_1, photons having energies E_3 and E_1 are emitted.

31. Which electron configuration would allow a hydrogen atom ( ) to absorb a photon of radiant energy, but not emit a photon?
(a) 1s
(b) 2s
(c) 3s
(d) none of these

32. The discovery of helium on the sun (before it was ( ) discovered on earth) involved the analysis of
(a) a chromatogram of solar vapor
(b) a sample of solar material
(c) atomic spectra
(d) lunar material

33. An atomic emission spectrum, like that discussed for ( ) hydrogen, may be produced in a sample of helium by
(a) bombarding it with high energy electrons
(b) heating it to sufficiently high temperatures
(c) causing an electric current to pass through the sample
(d) any of the above

34. Classical mechanics is expected to be inadequate in explain- ( ) ing the behavior of
(a) air at 25°C, 1 atm
(b) a ping pong ball
(c) the earth's rotation about the sun
(d) a nitrogen atom in a N_2 molecule

35. Wavelengths of electrons confined to atoms are the order of ( )
(a) 10^{-14} m
(b) 10^{-10} m
(c) 10^{-1} m
(d) 10 m

Problems

36. The electron energies in the hydrogen atom are given by the expression

$$E \text{ (joules)} = -2.18 \times 10^{-18}/n^2$$

For the transition n = 3 → n = 1, calculate
    (a) ΔE in joules
    (b) λ in nm (h = 6.63 × 10⁻³⁴ J · s, c = 3.00 × 10⁸ m/s, 1 nm = 10⁻⁹ m)

37. Consider an atom of titanium (atomic number 22).
    (a) Give the electron configuration.
    (b) Complete the following orbital diagram:

    1s   2s   2p   3s   3p   4s
    ( ) ( ) ( )( )( ) ( ) ( )( )( ) ( )

              3d
    ( )( )( )( )( )

    (c) How many unpaired electrons are there in the titanium atom?

38. Fill out the following table, giving an acceptable set of four quantum numbers for each electron in the oxygen atom (atomic number = 8).

Electron	n	ℓ	m_ℓ	m_s
First	___	___	___	___
Second	___	___	___	___
Third	___	___	___	___
Fourth	___	___	___	___
Fifth	___	___	___	___
Sixth	___	___	___	___
Seventh	___	___	___	___
Eighth	___	___	___	___

*39. Refer to Problem 6.10 in the text and calculate the wavelength of light just sufficiently energetic to ionize Li^{2+}. What part of the electromagnetic spectrum includes this wavelength?

*40. Calculate the temperature of H(g) at which the average translational energy would be just equal to the energy required to excite the atoms to their first excited state (n = 2).

SELF-TEST ANSWERS

1. **T** (The nucleus occupies a very small fraction of the atomic volume—Chapter 2.)
2. **F** (Shorter wavelength: $\Delta E = hc/\lambda$.)
3. **F** (Only if the atom is neutral.)
4. **T** (For ionic radii, see Chapter 8.)
5. **T**
6. **F** (It indicates the number of electrons in each orbital and their spins.)
7. **T** (Energy and frequency decrease, wavelength increases.)
8. **F** (Equal to but not less than.)
9. **T**
10. **T** (The force described in this chapter is the Coulomb electrostatic force.)
11. **c** (Coulomb force of attraction is much stronger than any gravitational or magnetic interaction.)
12. **d**
13. **c** (Corresponding to five geometric orientations.)
14. **c** ($\ell = 2$ means d orbitals, of which there are five, each capable of holding two electrons.)
15. **d** (Filled, it would be $4s^2 4p^6 4d^{10} 4f^{14}$; capacity = $2n^2$.)
16. **c** (Corresponding to the three orientations of the p orbitals: p_x, p_y, and p_z.)
17. **c** (With configuration $1s^2 2s^2 2p^3$.)
18. **d** (Eighteen electrons.)
19. **c** (Total of ten electrons, filling the lowest available levels.)
20. **b** (The 4s fills before the 3d; it must have the lower energy.)
21. **b** (Two electrons most easily lost are removed from 4s.)
22. **a** (Seven electrons in the neutral atom; NO_2^+ would contain $7 + 2(8) - 1 = 22$ electrons.)
23. **c** (Draw it out. Note that [Ar] is an abbreviation for argon's configuration.)
24. **d**
25. **a** (The electron is in the 5s orbital. For a systematic way of predicting configurations, see Chapter 7.)
26. **a** (2s fills before 2p.)
27. **c** (Recall Hund's rule.)
28. **d** (Second and third electrons both come from lower level, 1s.)
29. **a** (Again, a new, lower principal level is involved here, but not in Be or B.)
30. **d** (*The* photon energy would be $|\Delta E| = |E_1 - E_3|$.)
31. **a** (There is no lower energy state.)

32. c (That is, analysis of the light emitted by the sun.)
33. d
34. d (Object of small mass confined to a small region of space.)
35. b (As a first approximation, you might consider the atom to be one-dimensional with a width of 0.1 nm or 10^{-10} m. Then $\lambda = 2d/n = 2 \times 10^{-10}$ m. Or, use the Bohr velocity in $\lambda = h/mv$.)
36. (a) $E_3 = -0.242 \times 10^{-18}$ J, $E_1 = -2.18 \times 10^{-18}$ J
$\Delta E = E_1 - E_3 = -1.94 \times 10^{-18}$ J
(b) $\lambda = \dfrac{hc}{\Delta E} = \dfrac{6.63 \times 10^{-34} \times 3.00 \times 10^8 \text{ m}}{1.94 \times 10^{-18}} = 1.03 \times 10^{-7}$ m $= 103$ nm
37. (a) $1s^2 2s^2 2p^6 3s^2 3p^6 4s^2 3d^2$ or [Ar]$4s^2 3d^2$ (See Question 23 above.)
(b) 1s   2s   2p   3s   3p   4s
(↑↓) (↑↓) (↑↓)(↑↓)(↑↓) (↑↓) (↑↓)(↑↓)(↑↓) (↑↓)
            3d
(↑)(↑)( )( )( )
(c) 2

38.
	n	ℓ	m_ℓ	m_s
First	1	0	0	+1/2
Second	1	0	0	-1/2
Third	2	0	0	+1/2
Fourth	2	0	0	-1/2
Fifth	2	1	1	+1/2
Sixth	2	1	1	-1/2
Seventh	2	1	0	+1/2
Eighth	2	1	-1	+1/2

*39. $|\Delta E| = Z^2 B/n^2 = 9(2.179 \times 10^{-18}$ J$) = 1.961 \times 10^{-17}$ J

$E_{photon} = hc/\lambda$, $\lambda = \dfrac{(6.626 \times 10^{-34} \text{ J} \cdot \text{s})(2.998 \times 10^8 \text{ m/s})}{1.961 \times 10^{-17} \text{ J}} =$

1.013×10^{-8} m $= 10.13$ nm; ultraviolet

*40. $\Delta E = E_2 - E_1 = \left(\dfrac{1}{4}\right)(-2.18 \times 10^{-18}$ J$) - (-2.18 \times 10^{-18}$ J$)$
$= 1.63 \times 10^{-18}$ J/atom (see Problem 6.2 in the text)

For a mole of H: $\Delta E = 1.63 \times 10^{-18}$ J (6.02×10^{23})
$= 9.81 \times 10^5$ J/mol
Now, $\Delta E = E_{trans} = 3/2$ RT

$T = \dfrac{9.81 \times 10^5 \text{ J/mol}}{(3/2)[8.31 \text{ J/(mol} \cdot \text{K)}]} = 7.87 \times 10^4$ K $= 78\,700$ K!
(Also see Problem 6.39 in the text.)

SELECTED READINGS

Alternative discussions of electron structure are given in the following paperbacks:

Hochstrasser, R. M., *Behavior of Electrons in Atoms: Structure, Spectra, and Photochemistry of Atoms*, New York, W. A. Benjamin, 1964.

Pimentel, G. C., *Chemical Bonding Clarified through Quantum Mechanics*, San Francisco, Holden-Day, 1969.

Sisler, H. H., *Electronic Structure, Properties, and the Periodic Law*, New York, Van Nostrand, 1973.

Mainly of historical interest:

Bohr, N., On the Constitution of Atoms and Molecules, *Philosophical Magazine*, 26 (sixth series; July 1913), pp. 1–25.

Lagowski, J. J., *The Structure of Atoms*, Boston, Houghton Mifflin, 1964.

Lewis, G. N., *Valence and the Structure of Atoms and Molecules*, New York, Dover, 1966 (reprint).

For a look at the experimental support for configuration theory, and an extension to molecular structure:

Sanderson, R. T., Ionization Energy and Atomic Structure, *Chemistry* (May 1973), pp. 12–15.

Wahl, A. C., Chemistry by Computer, *Scientific American* (April 1970), pp. 54–66.

7
THE PERIODIC TABLE AND THE PROPERTIES OF ELEMENTS

QUESTIONS TO GUIDE YOUR STUDY

1. What properties of isolated atoms can be explained by electron configurations? Can this correlation be extended to the properties of bulk samples of matter?

2. How are elements related *to each other* in terms of their bulk properties? In terms of their configurations? (Can we predict many configurations by knowing just a few?) Are there similarities, or perhaps trends, in the observed properties of the chemical elements and their compounds?

3. How can this wealth of information be organized and made more manageable? What are some of the uses, besides organization of data, of the Periodic Table?

4. Is it possible to extend the Periodic Table to include, within or beyond its present content, yet other elements?

5. Are there exceptions and irregularities in the periodic classification? Are there properties of elements that are *not* periodic? If so, how are they explained?

6. In words, how might the organizing principle of chemical periodicity be stated?

7. Precisely what is meant by *periodicity* in the present context?

8. What are the sources of the elements used in commerce? What processes are used to separate them from their naturally occurring compounds? (Which elements occur uncombined?)

9. What are the limits to the natural reserves and how long will they last at the current rate of consumption? What are the alternatives to their exhaustion?

10. What are the ultimate sources of the elements? (What stellar processes, for example, lead to the observed distribution and relative abundances of the elements in the solar system? What geologic processes lead to the observed terrestrial distribution?)

11.

12.

YOU WILL NEED TO KNOW

Much information and several principles found in earlier as well as later chapters are organized in this one. Recognizing this, and learning the basic skills for using the Periodic Table (see below), will be very helpful in the remainder of your study of chemistry. Many of the basic ideas of preceding chapters are assumed here:

Concepts

1. How to define and work with density; how to distinguish between elements and compounds. It would be helpful to recognize more and more elements by their symbols and names—Chapter 1.

2. How to define and work with atomic number; AM—Chapter 2.

3. How to interpret and write chemical formulas and equations (in part, reviewed in this chapter)—Chapter 3.

4. How to write electron configurations. It would be very helpful to be able to do this, from memory, for elements 1 through 30—Chapter 6.

5. How to define ionization energy—Chapter 6.

Math

Essentially no math, and none that is new, is required in this chapter.

CHAPTER SUMMARY

The application of chemistry to the solution of practical problems (as well as those here and in the text) requires a familiarity with the actual properties of chemical substances. That familiarity approaches understanding when these properties can be correlated with and organized by a framework of theoretical principles. The quantity of chemical information now available and that being added each day is overwhelming. The organization of this data into a manageable system was first achieved by the discovery of the periodicity of observed properties of the elements and their compounds, over a century ago.

Only in the last five decades has this information been given a theoretical interpretation and unification. The periodicity of the properties of the elements is seen to be a result of the periodicity of outer-shell ("valence") electron configurations. (Or, if you wish, observed periodicity supports electron configuration theory.)

This correlation is simple: elements formerly grouped together on the basis of similar properties are now seen to be related by having the same number of electrons in the highest occupied principal energy level. For the A group ("representative") elements, this number is equal to the group number. All the 1A elements have one valence electron, all the 2A elements have two, and so on. We have already seen support for the idea that the outer-shell electrons determine chemical properties: thermochemical energies, resulting from the breaking and forming of bonds between atoms (Chapters 4 and 8), are of the same magnitude as the energies required for the removal of these particular electrons (ionization energies, Chapter 6).

By grouping together elements with similar properties, we can study their chemistry as a group, rather than deal with each of them separately. The elements sulfur, selenium, and tellurium are more alike (e.g., in their first ionization energies) than are the elements silicon, phosphorus, and sulfur. As a general rule, properties show smaller and more gradual changes within a group than within a period. (Interestingly, the elements 1 through 9 are not so very representative of their respective groups, a result of their specially small size.)

The Periodic Table is a valuable device for remembering trends in properties. We find, for example, that as we read from left to right atomic radius and metallic character both decrease while ionization energy increases; as we consider elements of higher atomic number within a group, reading down a column, atomic radius and metallic character increase while ionization energy decreases. All these trends, and many, many others, can be explained in terms of electronic structure. (Other properties include, to name a few: ΔH_f of compounds; acidic and basic character of oxides of the elements; melting and boiling points; density; crystal structures.) We see, for

92 • 7–The Periodic Table and The Properties of Elements

example, ionization energy increasing within a period to be a result of an increasing nuclear charge, with little additional shielding to diminish its magnitude.

Again, to emphasize the usefulness of the Periodic Table, the chemistry student and the research chemist alike are able to predict properties not available or not yet determined. As this is being written, evidence is reported for the existence of element 126; the observed properties correspond to those predicted.

The study of the periodic classification allows you to bring order and deeper meaning to the detailed information of the previous chapters as well as those which follow. Make an effort to understand the correlation of properties with electron structure so that the descriptive chemistry that follows will not be just one surprise after another.

BASIC SKILLS

1. Given the position in the Periodic Table of an A subgroup element, state its outer electron configuration.

See Example 7.1 and the discussion that immediately precedes it. This skill is required to work Problems 7.1, 7.7, and 7.22.

2. Predict the physical properties of an element, given those of elements surrounding it in the Periodic Table.

See Example 7.2 and Problems 7.2, 7.13, and 7.27. Note that the predicted values are only approximate; if sufficient data are available a more reliable estimate can be made by graphical interpolation, as illustrated in Problem 7.13.

3. Use the Periodic Table to predict relative values of the following properties of elements: atomic radius, electronegativity, ionization energy, and metallic character.

A typical prediction of this type is shown in Example 7.3. Problems 7.3, 7.9, and 7.24 give you a chance to exercise this skill. In predicting how properties vary with position in the Periodic Table you may find the following diagram helpful; properties are nearly constant along the diagonal line D.

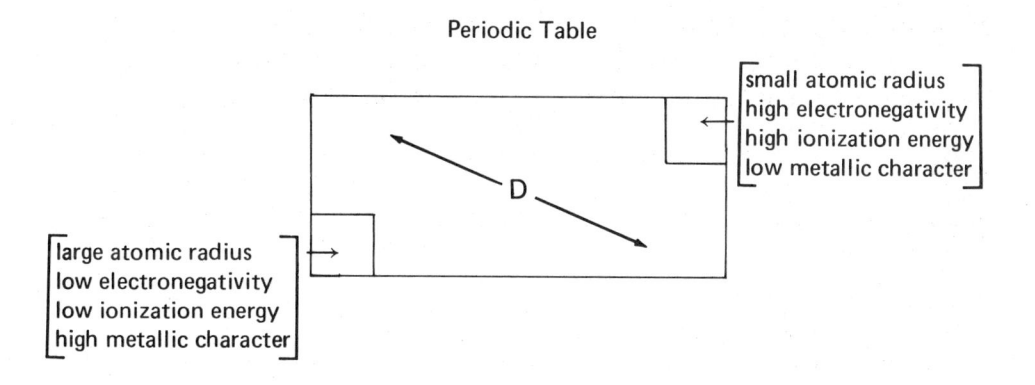

Periodic Table

[large atomic radius / low electronegativity / low ionization energy / high metallic character]

[small atomic radius / high electronegativity / high ionization energy / low metallic character]

D

4. Use the Periodic Table to predict the formulas of binary or ternary compounds, given those of analogous compounds of elements in the same groups.

See Example 7.4 and Problems 7.4, 7.14, and 7.29. Students, in attempting to work problems of this type, are sometimes intimidated by the nomenclature of the compounds involved. One general hint: the suffix "ate" implies the presence of oxygen as well as one other element in an anion. Thus, the sulfate, selenate, and arsenate anions all contain oxygen in addition to S, Se, and As, respectively. You may also find helpful the general rules for naming inorganic compounds given in Appendix 3.

In addition to these general skills, Chapter 7 contains a considerable amount of descriptive chemistry with which you should become familiar. In particular, you should acquire a general knowledge of:

— the chemical and physical properties of the 1A elements (alkali metals) and 7A elements (halogens).
— the manner in which various metals (1A, 2A, and transition metals) react with water and with halogens.
— the properties that are characteristic of transition and inner transition metals, as distinguished from those in the A subgroups.
— the principal ores or sources of elements as related to their positions in the Periodic Table.
— the chemical reactions commonly used to obtain metals from their ores.

This is the first chapter in the text which is primarily descriptive rather than quantitative; you will note that virtually no calculations are called for in the problems. This "change of pace" often shakes up students who have been relying on mathematical skills up to this point. There are principles of descriptive chemistry such as those illustrated in the basic skills listed above.

94 • 7–The Periodic Table and The Properties of Elements

We must admit, however, that a considerable amount of memorization will be necessary if you are to become proficient in this area. Like it or not, chemistry never has been, is not now, and never will be an exact science.

SELF-TEST

True or False

1. The earliest Periodic Table was based on electron configurations of atoms. ( )

2. The law of chemical periodicity might be stated as: properties of the elements are a periodic function of atomic mass. ( )

3. A vertical column in the Periodic Table is known as a *group* or *family*. ( )

4. The number of elements belonging to the sixth period is 18. ( )

5. Since density = mass/volume, cadmium (AM = 112) is expected to be twice as dense as iron (AM = 56). ( )

6. The properties of element 107 should be similar to those of any Group 7B element. ( )

7. The properties of the element with atomic number 116, should it be observed, will be unpredictable. ( )

8. The transition elements exhibit intermediate electrical conductivity, boiling point, ... ( )

9. Within a given period of the representative elements, s and p orbitals fill from left to right. ( )

10. For any atom, the second ionization energy is larger than the first. ( )

Multiple Choice

11. The element which immediately follows lanthanum is ( )
    (a) hafnium   (b) cerium
    (c) thorium   (d) undiscovered

12. The group number for the A-group element gives the ( )
    (a) effective nuclear charge
    (b) number of valence electrons
    (c) number of elements in the group
    (d) value of n for the valence electrons

13. Within a given period of number n, ( )
    (a) n electrons occupy the highest principal energy level
    (b) n is the effective nuclear charge
    (c) n is the principal energy level with d orbitals being filled
    (d) n is the principal energy level with s and p orbitals being filled

14. Compounds of bromine and any Group 1A metal are ( )
expected to have the formula
    (a) $MBr_3(s)$ (b) $MBr_2(s)$
    (c) $MBr(s)$ (d) $MBr(l)$

15. Per mole of calcium salt, the largest mass of calcium might ( )
be recovered from
    (a) CaF_2
    (b) $CaCl_2$
    (c) $CaBr_2$
    (d) all contain the same amount

16. The molar enthalpies of formation of HF, HCl, and HBr are ( )
−268.6, −92.3, and −36.2 kJ. That of HI is likely to be
    (a) more negative than −268.6
    (b) −132.4, the average
    (c) more positive than −36.2
    (d) unpredictable

17. Of the halogens, the one reacting most readily with ( )
hydrogen is fluorine. That which reacts least readily is likely to be
    (a) Cl_2 (b) Br_2
    (c) I_2 (d) At_2

18. The elements characterized as nonmetals are situated in the ( )
Periodic Table at the
    (a) far left   (b) center
    (c) bottom   (d) top right

19. An example of a metalloid is ( )
    (a) $_{12}Mg$ (b) $_{14}Si$
    (c) $_{16}S$ (d) $_{18}Ar$

20. To indicate that an atom has ten electrons in its third ( )
principal energy level, you would write
    (a) 3^{10} (b) $3d^{10}$
    (c) $3s^2 3p^6 3d^2$ (d) $3s^2 3p^6 4s^2$

21. The electron configuration of gallium is expected to be (use ( )
only the Periodic Table!)

(a) [Ar]$4s^2 3d^{10} 4p^1$ (b) [Ar]$4s^1 3d^5 4p^6$
(c) [Ar]$4s^2 4p^1$ (d) $1s^2 2s^2 2p^6 3s^2 3p^1$

22. The orbital diagram for any 6A element will show   ( )
    (a) six electrons
    (b) six unpaired electrons
    (c) two half-filled orbitals
    (d) filled p orbitals

23. Ionization of a fourth period transition series atom to give a ( ) +1 ion involves the removal of an electron from the orbital
    (a) 3s   (b) 3p
    (c) 3d   (d) 4s

24. For the given configurations, the first ionization energy is ( ) probably smallest for the atom with the configuration
    (a) $ns^2 np^3$ (b) $ns^2 np^4$
    (c) $ns^2 np^5$ (d) $ns^2 np^6$

25. In Group 8A, the first ionization energy decreases with ( ) increasing atomic number. This is correlated with and explained, at least in part, by
    (a) increasing atomic mass
    (b) increasing nuclear charge
    (c) increasing atomic radius
    (d) increasing boiling point

26. The formation of a +2 ion is probably easiest for   ( )
    (a) $_{11}$Na   (b) $_{12}$Mg
    (c) $_{17}$Cl   (d) $_{20}$Ca

27. In which of the following series are the atoms arranged in ( ) order of increasing ionization energy?
    (a) Li, Na, K   (b) B, Be, Li
    (c) O, F, Ne   (d) C, P, Se

28. Atomic radius decreases   ( )
    (a) within a family, from low to high atomic number
    (b) within a period, from low to high atomic number
    (c) as electrons enter higher principal energy levels
    (d) with more effective shielding of nuclear charge

29. Which of the following species would you expect to have ( ) the largest radius?
    (a) $_{11}$Na$^+$ (b) $_{10}$Ne
    (c) $_9$F$^-$ (d) $_8$O^{2-}

30. If each of the following solids consisted of the same ( ) compact arrangement of atoms, which one would occupy the largest volume per mole?

(a) $_{19}$K (b) $_{37}$Rb
(c) $_{38}$Sr (d) $_{48}$Cd

31. Of the elements listed below, that which is most electro- ( )
negative is
(a) $_8$O (b) $_{13}$Al
(c) $_{15}$P (d) $_{16}$S

32. Within a group, say 6A, metallic character ( )
 (a) increases with increasing atomic number
 (b) decreases with increasing atomic number
 (c) remains more or less constant throughout
 (d) increases with increasing ionization energy

33. Elements known to ancient civilizations are likely to have ( )
included all of the following except
 (a) Ag (b) Hg
 (c) S (d) Na

34. Which one of the following properties of an element would ( )
you guess *not* to be a periodic function of atomic number?
 (a) atomic volume (b) specific heat
 (c) ionization energy (d) boiling point

35. To estimate the density of hafnium, $_{72}$Hf, one would ( )
expect to average the densities of
 (a) La and Ta (b) Zr and Ce
 (c) Zr and U (d) Lu and Ta

36. The atomic number of the next-to-be-discovered noble gas, ( )
below radon, Rn, would be
 (a) 109 (b) 118
 (c) 173 (d) 222

37. Which of the following elements do you expect $_{30}$Zn to ( )
resemble most closely in its chemical properties?
 (a) $_{20}$Ca (b) $_{21}$Sc
 (c) $_{31}$Ga (d) $_{48}$Cd

38. Sulfide ores are among the more important sources of ( )
 (a) noble gases (b) alkali metals
 (c) metalloids (d) transition metals

39. The process in which a metal sulfide is converted to the ( )
metal or an oxide by heating in air is called
 (a) combustion (b) roasting
 (c) flotation (d) oxidation

40. At the current rate of growth of consumption, it seems ( )
likely that in less than a century we will exhaust

7—The Periodic Table and The Properties of Elements

  (a) the rare earths
  (b) the platinum metals
  (c) most commercial metals
  (d) metals potentially available from low-grade ores

Problems

41. In a different universe, a Periodic Table has been developed, based on the chemical properties of the elements found there. A portion of the table is reproduced below (atomic mass appears below the symbol for the element).

φ	O	θ
20	25	30
Б	Pv	Г
60		80
Д	Щ	З
100		130

The formulas of the chlorides of Б and З are known to be БCl₃ and ЗCl₅. For the element Pv, predict
  (a) the atomic mass
  (b) the formula of its chloride
  (c) the element in our periodic chart which most closely resembles Pv.

42. Given the formulas $KMnO_4$, $BaWO_4$, and $NaBiO_3$, predict the formulas of the ternary compounds formed by
  (a) Cs, As, and O (b) Ba, Re, and O
  (c) Mg, Mo, and S

43. Write balanced equations for the reactions that take place when
  (a) potassium metal is added to water
  (b) barium chloride is electrolyzed to yield the free elements
  (c) cadium sulfide, CdS, is roasted in air to form cadmium oxide and a gas with a choking odor

*44. Suppose the annual growth of aluminum consumption continues at 6.4%. In what year would we consume *twice* the 1970 amount?

*45. Use a library to find the following information about chlorine:
  (a) the annual U.S. production, in billions of kilograms, in the years 1960, 1968, and 1970 to date. Plot a smooth curve of

production versus year. Estimate the 1980 production.
(b) the source(s) of Cl_2 and its (their) estimated reserves.

SELF-TEST ANSWERS

1. **F** (Rather, on experimental observations of regularities in chemical and physical properties.)
2. **F** (Atomic number, not mass as was originally thought. Besides Te and I, can you find other sequences of elements which would be inverted if periodicity did depend on mass?)
3. **T**
4. **F** (32, counting the lanthanides.)
5. **F** (Atoms may pack in different ways, one kind filling a given volume more completely than another; see Chapter 11. Observed densities are actually 8.6 and 7.9 g/cm^3, respectively. Comparisons of density are more accurately made within a period or a group.)
6. **T** (107 should be placed directly below rhenium.)
7. **F** (Precisely by comparison with prediction, evidence already suggests its existence; see *Chemical and Engineering News* (June 21, 1976), p. 6.)
8. **F** (High, not intermediate, values for these two properties.)
9. **T** (From ns^1 to $ns^2 np^6$ for period and principal energy level n.)
10. **T** (It is yet more difficult to remove an electron from a cation than from a neutral atom.)
11. **b**
12. **b** (Valence = outer shell = highest principal energy level. Choice a is not far from the truth.)
13. **d** (And $(n-1)$ d orbitals, starting with n = 4.)
14. **c** (Analogous to the familiar NaCl; a solid is expected.)
15. **d** (Each would contain one mole of Ca. Review Chapter 3.)
16. **c** (The trend is in this direction. Many, many properties can be correlated with the position of an element in the Periodic Table. Most tabulations of data throughout the text can be given additional meaning in this way.)
17. **d**
18. **d**
19. **b**
20. **c** (All ten belong to n = 3. Rules for writing configurations are in Chapter 6.)
21. **a** (Thirty-one electrons total; 3A implies $ns^2 np^1$; fourth period, n = 4.)
22. **c** (Corresponding to $ns^2 np^4$.)

23. d (4s is filled *and emptied* before 3d, 5s before 4d ... ; see Chapters 6 and 8. Experimental support is given in Chapter 21.)
24. b (The overall trend, from ns^2np^3 to ns^2np^6, is toward higher ionization energy as a result of larger nuclear charge. But why is it easier to remove an electron from ns^2np^4 than from ns^2np^3, or from ns^2np^1 than ns^2? Recall the special stability of half- and completely filled orbitals, Chapter 6.)
25. c (Along with greater shielding of the nuclear charge.)
26. d (Both Mg and Ca, with ns^2 configurations, can easily lose two electrons; Ca more so, for the reasons given in Question 25, above.)
27. c (C, P, and Se, on a diagonal through the table, have very similar ionization energies.)
28. b (Larger nuclear charge, with little additional shielding, contracts electron cloud.)
29. d (Of these isoelectronic species, O has the smallest nuclear charge.)
30. b
31. a (Electronegativity increases toward the top and the right.)
32. a (Increasing toward the bottom and the left.)
33. d (Highly reactive, hence difficult to isolate from its compounds.)
34. b (That of the metals is, more or less, uniformly different from that of the nonmetals. Recall Law of Dulong and Petit, Chapter 2.)
35. d (Elements 71 and 73.)
36. b
37. d
38. c
39. b
40. c (Which also includes b; see Table 7.4 in the text.)
41. (a) 70 (b) $PvCl_4$ (c) probably Ge
42. (a) $CsAsO_3$ (b) $Ba(ReO_4)_2$ (c) $MgMoS_4$
43. (a) $K(s) + H_2O(l) \rightarrow \frac{1}{2} H_2(g) + KOH(s)$

  (b) $BaCl_2(s) \rightarrow Ba(s) + Cl_2(g)$

  (c) $CdS(s) + \frac{3}{2} O_2(g) \rightarrow CdO(s) + SO_2(g)$

*44. One way to answer this would be graphically, by extrapolation. Another considers the exponential, or logarithmic, relationship $\log \frac{x_0}{x} = \frac{kt}{2.30}$, where x_0 would be the consumption in year t = 0, x would be the consumption in year t, and k is a constant of proportionality. (See, e.g., Chapter 16 as well as readings listed below.) To find k, consider the consumption for 1970 (year t = 0)

and 1971 (t = 1):
$$\log \frac{x_0}{1.064x_0} = \frac{k(1)}{2.30}, k = -0.0620 \text{ a}^{-1}$$
Now, to find t for $x = 2x_0$:
$$\log \frac{x_0}{2x_0} = \frac{-0.0620t}{2.30}, t = 11 \text{ a; hence, } 1981$$

*45. Is there exponential growth? Does it begin to level off? Are the resources renewable? References might include annual reviews of the chemical industry in *Chemical and Engineering News*, publications of the Bureau of the Census, etc. For 1975, some 36×10^9 kg Cl_2 were produced, *down* 16% from 1974 production.

SELECTED READINGS

Chemical periodicity is also discussed in:

Puddephatt, R. J., *The Periodic Table of the Elements*, New York, Oxford University Press, 1972.

Sanderson, R. T., *Chemical Periodicity*, New York, Reinhold, 1960.

Seaborg, G. T., Prospects for Further Considerable Extension of the Periodic Table, *Journal of Chemical Education* (October 1969), pp. 626-634.

Sisler, H. H., *Electronic Structure, Properties, and the Periodic Law*, New York, Van Nostrand, 1973.

For discussions of the discovery and properties of specific elements or families:

Chedd, G., *Half-Way Elements: The Technology of Metalloids*, Garden City, N. J., Doubleday, 1969.

Handbooks of Chemistry, listed in the Preface.

Rochow, E. G., *The Metalloids*, Boston, D. C. Heath, 1966.

Rochow, E. G., *Modern Discriptive Chemistry*, Philadelphia, W. B. Saunders, 1977.

Weeks, M. E., *Discovery of the Elements*, Easton, Pa., Journal of Chemical Education, 1968.

Elemental abundance and availability is discussed in:

Bachmann, H. G., The Origin of Ores, *Scientific American* (June 1960), pp. 146-156.

On the exponential growth of science and technology, on the limits to growth:

Lapp, R. E., *The Logarithmic Century*, Englewood Cliffs, N. J., Prentice-Hall, 1973.

Meadows, D. H., *The Limits to Growth*, New York, Universe, 1972.

Price, D. J. de S., *Little Science, Big Science*, New York, Columbia, 1963.

Skinner, B. J., A Second Iron Age Ahead?, *American Scientist* (May-June 1976), pp. 258-269.

8
CHEMICAL BONDING

QUESTIONS TO GUIDE YOUR STUDY

1. Why do bonds form between neutral atoms, as well as between ions? (Why don't all the atoms in the universe bond together in one super molecule?)

2. Is there anything common to all "types" of bonds, whether ionic, covalent, metallic, etc.?

3. How do you account for the observed differences in the strengths of bonds?

4. How are the size and shape of a molecule related to the electronic structures of the component atoms? Is there a simple, reliable way of predicting the geometry?

5. Can you predict formulas of compounds from a theoretical basis? In particular, in terms of the electronic structures of the atoms involved?

6. How does bonding theory help us explain such properties of materials as color, electrical conductivity, boiling point, and magnetic behavior?

7. How would you experimentally determine bond energies, lengths, angles, and polarities?

8. How would you account for the fact that an atom may bond to just one other atom, sometimes to two others, or three ... (e.g., C in CO, CO_2, CH_4)? Is the number of bonds predictable?

9. You have "constructed" individual atoms by filling atomic orbitals. Can you construct a molecule in an analogous manner?

10. Can you now give a more detailed description as to what occurs at the atomic-molecular level when a chemical reaction takes place?

11.

12.

YOU WILL NEED TO KNOW

Concepts

1. How to write electronic structures for representative elements—neutral atoms and also monatomic ions (say, for elements of atomic number 1 to 30, and others by analogy)—Chapters 6, 7.

2. This means that you need to be able to use the Periodic Table as a correlating tool (example: the "valence electron" structure for $_{83}$Bi is analogous to that of Sb and the other elements in Group 5A of the table: Bi—$6s^2 6p^3$; Sb—$5s^2 5p^3$)—Chapter 7.

3. The shapes and relative sizes of atomic orbitals—Chapter 6.

4. How to apply the principle of electroneutrality to determine formulas of ionic compounds, as well as how generally to interpret formulas—Chapter 3.

Math

1. How to calculate bond energies from other heats (enthalpies) of reaction, and vice versa—Chapter 4.

2. A familiarity with the geometry (symmetry and angles) of several plane and solid figures, particularly the tetrahedron, would be helpful—see a math text or the Readings of Chapter 11 and the Preface. A molecular model set would also be very useful in visualizing the geometric consequences of bonding.

CHAPTER SUMMARY

Most familiar materials consist *not* of individual isolated atoms (the subject of Chapter 6), but of aggregates of atoms and ions. The existence of molecular clusters of atoms as well as the existence of condensed states supports the idea that attractive forces, or "bonds," exist between atoms, ions, and even molecules. It is convenient, though somewhat arbitrary, to distinguish two kinds of forces: the generally stronger chemical bonds (this chapter) and the usually weaker intermolecular forces—or, perhaps, "physical bonds" (next chapter).

All of these attractive forces are electrostatic in nature. Their magnitudes vary from the very weak attraction between neighboring atoms in liquid helium, where 0.1 kJ is enough energy to separate a mole of the "bonded" particles into a monatomic gas, to the very strong bond between atoms in a molecule of carbon monoxide, with a bond energy of 1080 kJ/mol, and almost anything in between. (Further examples of the range of attractive forces: $Na_2(g)$ has a bond energy of 75 kJ/mol; HF(g), 560 kJ; for a hydrogen bond, the bond energy is typically between 4 and 40 kJ/mol.)

Again, it is convenient, as well as somewhat arbitrary, to classify chemical bonds according to the extent to which the electron distribution within atoms is changed when the atoms are strongly attracted to one another. The chemical bonds discussed in this chapter are of two types:

1. *Ionic bonds* between positive and negative ions. Ions are formed by the transfer of electrons from an element of low electronegativity (a metal) to an element of high electronegativity (a nonmetal). When two elements differ in electronegativity by more than 1.7 units, we expect the bonding between them to be predominantly ionic. We can readily deduce the charges of ions having a noble gas structure: 1A, +1; 2A, +2; Al and 3B, +3; 6A, −2; 7A, −1. The transition metals commonly form cations with charges of +2 (Zn^{2+}, Cu^{2+}, Ni^{2+}, Fe^{2+}, Co^{2+}), +3 (Cr^{3+}, Fe^{3+}, Co^{3+}), or less frequently +1 (Ag^+). Many of the most common anions are polyatomic (OH^-, NO_3^-, SO_4^{2-}, CO_3^{2-}, PO_4^{3-}); the NH_4^+ ion is one of few polyatomic cations. Formulas of ionic compounds are readily deduced by imposing the condition of electroneutrality (e.g., $Zn(OH)_2$, $Cr(NO_3)_3$, Ag_2SO_4).

2. *Covalent bonds*, which consist of a pair of electrons shared between two atoms. We can expect covalent bonds to be formed when a nonmetal combines with another element which differs from it in electronegativity by less than 1.7 units. The greater the difference in electronegativity, the more polar will be the bond. In H_2 and Cl_2 where $\Delta EN = 0$, the covalent bond is nonpolar; i.e., the electron pair is equally shared by the bonded atoms. In contrast, the bond in the HCl molecule is polar—the bonding electrons are displaced toward the more electronegative Cl atom. The displacement of electrons that characterizes a polar covalent bond strengthens and shortens it; the formation of double bonds (two pairs of electrons between bonded atoms) or triple bonds (three pairs of electrons) has the same effect.

A major portion of this chapter is devoted to the structure of molecules and polyatomic ions, both of which are held together by covalent bonds. Here we are primarily interested in two factors: the way in which the valence electrons are distributed and the angles between the covalent bonds. The electron distributions in a wide variety of polyatomic ions and molecules can be represented satisfactorily by Lewis structures; these in turn can be derived by following the simple rules listed in Section 8.4.

Once you have arrived at the Lewis structure for a covalently bonded species, it is relatively easy to predict the bond angles and hence the geometry of the molecule or polyatomic ion. To do this, we apply the simple principle that electron pairs surrounding an atom are oriented to be as far apart as possible (so as to minimize their mutual repulsion). This implies, for example, that four pairs of electrons around an atom will be directed toward the corners of a regular tetrahedron. The principle is readily extended to molecules containing multiple bonds by noting that, so far as geometry is concerned, a multiple bond behaves as if it were a single electron pair.

Knowing the geometry of a molecule and the electronegativities of its atoms, we can decide whether it is polar or nonpolar. Molecules in which polar bonds are arranged unsymmetrically with respect to one another (e.g., H_2O, NH_3) are polar. Molecules in which all the bonds are nonpolar (e.g., H_2, Cl_2) or in which all the polar bonds "cancel" one another by their symmetrical arrangement (e.g., CO_2, CCl_4) are nonpolar.

The concepts that we have just reviewed are fundamental to an understanding of chemical bonding and molecular structure. There are certain other ideas with which you should be familiar. These include "resonance" and "hybridization" which were grafted onto the atomic orbital approach to rationalize some of its shortcomings. (Resonance structures constitute, in essence, an admission that we cannot represent a particular species in terms of one simple Lewis-type picture: the species has only one structure, and the problem is in finding a satisfactory way of describing it!)

Finally, there is included in this chapter an introduction to a somewhat more fundamental approach to molecular structure, the molecular orbital method. If you go on to take further courses in chemistry, you will hear a great deal more about molecular orbitals; in this chapter, we have done little more than outline the basic approach and apply it to some very simple molecules.

BASIC SKILLS

1. With the aid of a Periodic Table, derive the formulas of ionic compounds.

This skill is illustrated in Example 8.1. Note that in order to apply it you must:

— be familiar with the way in which ionic compounds are named (see the discussion in the text preceding Example 8.1 and the more extensive discussion in Appendix 3).

Basic Skills • 107

— be able to predict the charges of monatomic ions of the A-group elements (Table 8.1).
— know the charges of some of the more common cations formed by the transition and post-transition metals (Table 8.2).
— be familiar with the charges and names of the common polyatomic ions (Table 8.3).

To gain practice in this skill, which will be useful in later chapters as well as in this one, try Problems 8.1, 8.9, 8.10, 8.28, and 8.29.

2. Given a Periodic Table, write electron configurations for monatomic ions.

For ions of elements in Groups 1A, 2A, 3B, 6A, and 7A this poses no problem, since they have the configuration of the nearest noble gas. For the ions of transition metals to the right of Group 3B you must realize that, unlike the situation in the corresponding atoms, d electrons are lower in energy than the s electrons in the next principal level (e.g., 3d is lower in energy than 4s).

What is the electron configuration of K^+? _____ of Cr^{2+}? _____
The K^+ ion will have the electron configuration of the argon atom (Chapter 6): $1s^2 2s^2 2p^6 3s^2 3p^6$. In the Cr^{2+} ion, there are 22 electrons (two less than the 24 of the neutral Cr atom). The four electrons beyond the argon configuration will be located in the 3d level: $1s^2 2s^2 2p^6 3s^2 3p^6 3d^4$ or $[Ar]3d^4$.

See Problems 8.11 and 8.30. Note that you should also be able to write orbital diagrams for ions, using the principles discussed for atoms in Chapter 6.

3. Given a Periodic Table (or a table of electronegativities), compare the relative degrees of polarity of different bonds.

See Example 8.2 and Problem 8.2, where only a Periodic Table is required. In Problems 8.12 and 8.31, a table of electronegativities is needed (Fig. 7.4, Chapter 7).

4. Draw Lewis structures for molecules and polyatomic ions.

This skill is shown in Example 8.3; it is used directly to solve Problems 8.3, 8.13, 8.14, 8.32, and 8.33. Problems 8.17 and 8.36 are slightly more

108 • 8–Chemical Bonding

subtle (count the number of valence electrons), but the same principle is involved.

We cannot emphasize too strongly that the ability to draw Lewis structures is fundamental to a great many other skills. You cannot predict the geometry of a molecule, decide whether it is polar or nonpolar, write resonance forms, indicate the type of hybrid bonds present, and so forth, unless you know its Lewis structure.

Frequently, more than one plausible Lewis structure can be written for a molecule or polyatomic ion. Such structures ordinarily differ only in their skeletons — i.e., the order in which atoms are bonded to one another. This point is discussed in Example 8.3. In part (b), the correct structure of the sulfate ion is deduced by applying the rule that, in oxyanions, all the oxygen atoms are ordinarily bonded to the central, nonmetal atom. Again, in part (c), we arrive at the correct structure for formaldehyde by taking into account the fact that carbon generally forms four bonds. As you progress in your study of chemistry, you will gradually add to your store of rules of this type.

5. Given or having derived the Lewis structure of a molecule or polyatomic ion, deduce its geometry.

The reasoning here is shown in Example 8.4 (no multiple bonds) and Example 8.5 (species containing double bonds). There are only two principles involved:

a. The electron pairs around an atom tend to be as far apart as possible. Thus, two electron pairs are oriented at 180° to each other; three electron pairs are directed toward the corners of an equilateral triangle; four electron pairs are directed toward the corners of a regular tetrahedron.

b. So far as geometry is concerned, a multiple bond behaves as if it were a single bond. Thus, in SO_2, we ignore the "extra" electron pair in the double bond and pretend that there are only three pairs of electrons around sulfur to arrive at a 120° bond angle.

Problems 8.4, 8.18, 8.19, 8.20, 8.37, 8.38, and 8.39 give you a chance to apply these principles. Notice that, as pointed out previously, you need to know the Lewis structure before you can predict the geometry.

One point here sometimes causes confusion. In describing molecular geometry, we indicate only the positions of the atoms; the location of unshared electron pairs is not included. Thus we would not ordinarily refer to the NH_3 molecule as a "tetrahedron." Instead, we would describe only the positions of the three H atoms around the nitrogen, perhaps by calling it a "tetrahedron with one corner missing" or, better, a "pyramidal" molecule (i.e., a triangular pyramid with three H atoms and one N atom at the corners).

6. **Given or having derived the geometry of a molecule or polyatomic ion, predict whether it will be a dipole.**

The general principle here is a simple one; a molecule can be nonpolar only if

a. all the bonds are nonpolar, as in H_2 or P_4.
b. the various polar bonds "cancel" each other, as would be the case in the symmetrical molecules

$$X-Y-X \quad \text{or} \quad \begin{array}{c} X \diagdown \quad \diagup X \\ Y \\ | \\ X \end{array}$$

If the symmetry is destroyed by substituting an atom other than X as in

$$Z-Y-X \quad \text{or} \quad \begin{array}{c} Z \diagdown \quad \diagup X \\ Y \\ | \\ X \end{array}$$

or by changing the bond angle as in

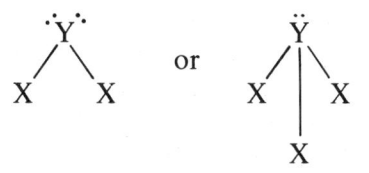

$$\begin{array}{c} \ddot{Y} \\ \diagup \diagdown \\ X \quad X \end{array} \quad \text{or} \quad \begin{array}{c} \ddot{Y} \\ \diagup | \diagdown \\ X \quad X \\ | \\ X \end{array}$$

then the molecule is polar. This principle is illustrated in Example 8.6 and in Problems 8.5, 8.21, 8.22, 8.40, and 8.41. Notice particularly in the example that you ordinarily need to know the geometry before you can predict polarity; geometry in turn can be deduced if the Lewis structure is known.

7. **Given the Lewis structure of a molecule, give the hybridization around the central atom.**

The hybrid orbitals discussed here include sp (two pairs of electrons), sp^2 (three pairs of electrons), and sp^3 (four pairs of electrons). Note that only one pair of electrons in a multiple bond can be hybridized: the other pair(s) are not hybridized. Notice also the correlation between hybridization and geometry (Table 8.7); if the geometry of a species is known, its hybridization can be predicted.

110 • 8–Chemical Bonding

This skill is illustrated in Example 8.7 and required to work Problems 8.6, 8.23, 8.24, 8.42, and 8.43.

8. **Write molecular orbital diagrams for simple diatomic species.**

The principles here are entirely analogous to those used in Chapter 6 to write atomic orbital diagrams for isolated atoms. Each MO, like each AO, can hold two electrons. When two orbitals of equal energy are available, electrons tend to enter singly with parallel spins (note, for example, the structures of B_2 and O_2 in Table 8.8). Unfortunately, to work out MO diagrams, you must learn a new set of symbols (e.g., σ_{2s}^b, π_{2p}^*, ..) and a new filling order. The latter can be deduced from Table 8.8.

Example 8.8 shows how an MO diagram can be derived for a simple diatomic species. Problems 8.7, 8.25, 8.26, 8.44, and 8.45 test this skill further and illustrate some of the applications of MO structures.

As you can judge from the number of skills listed (we've left out a few of the less vital ones), there is a lot of material in this chapter. This may explain why students often find it one of the more difficult chapters to master. The principles themselves are relatively simple; if you devote sufficient time to this material, you should have no trouble with it.

SELF-TEST

True or False

1. Chemical bonds form because one or more electrons are ( ) attracted simultaneously by two or more nuclei.

2. Chemical bonds almost never form unless half-filled orbitals ( ) are available.

3. A reasonable formula for a compound of aluminum and ( ) bromine would be $Al_3 Br$.

4. Bond energy increases in the order: single bond < double ( ) bond < triple bond.

5. Bond length increases as bond energy increases. ( )

6. The coefficient of boron, $_5B$, in the equation for the ( ) reaction of boron with oxygen, can be predicted from the electronic structures of boron and of oxygen.

7. A binary compound consisting of an element having a low ( ) ionization potential and a second element having a high electronegativity is likely to possess covalent bonds.

8. Experimental results support the idea that a certain ( )
molecule, AB$_2$, is linear (that is, all three nuclei lie along a straight
line). This must mean that there are no unshared electrons on the
central atom.

9. There are 24 valence electrons in the SO$_3{}^{2-}$ ion. ( )

10. Of the species O$^+$, O, and O^{2-}, the largest is O^{2-} and the ( )
smallest is O$^+$.

Multiple Choice

11. The ion, Ni^{2+}, would have the electron configuration ( )
    (a) [Ar]3d^{8}4s^2 (b) [Ar]3d^8
    (c) [Ar]3d^{7}4s   (d) [Ar]3d^{6}4s^2

12. Of the elements listed below, which would you expect to ( )
have the largest attraction for bonding electrons?
    (a) $_3$Li   (b) $_{13}$Al
    (c) $_{26}$Fe   (d) $_{16}$S

13. The forces most suited to account for the fact that atoms ( )
often combine to form molecules are
    (a) nuclear   (b) magnetic
    (c) electrical   (d) polar

14. Which one of the following contains both ionic and covalent ( )
bonds?
    (a) NaOH   (b) HOH
    (c) C$_6$H$_5$Cl   (d) SiO$_2$

15. For the reactions ( )

$$CH_4(g) \rightarrow C(g) + 4\ H(g), \Delta H = 1656\ kJ$$
$$H_2C = CH_2(g) \rightarrow 2\ C(g) + 4\ H(g), \Delta H = 2254\ kJ$$

The C=C bond energy, in kJ/mol, is
    (a) −598   (b) +414
    (c) +598   (d) +711

16. Which one of the following species would have an unpaired ( )
electron?
    (a) SO$_2$ (b) NO$_2{}^-$
    (c) NO$_2{}^+$ (d) NO$_2$

17. Which of the following bonds would be the least polar? ( )
    (a) H−F   (b) O−F
    (c) Cl−F   (d) Ca−F

112 • 8–Chemical Bonding

18. The fact that all bonds in SO_3 are the same is accounted for by ( )
    - (a) the idea of resonance
    - (b) the octet rule
    - (c) electron repulsion theory
    - (d) knowing that sulfur always forms three equivalent bonds

19. Indicate which one of the following does not obey the octet rule: ( )
    - (a) NO_3^-
    - (b) O_3
    - (c) HCN
    - (d) BeF_2

20. The fact that the BeF_2 molecule is linear implies that the Be—F bonds involve ( )
    - (a) sp hybrids
    - (b) sp^2 hybrids
    - (c) sp^3 hybrids
    - (d) resonance

21. The molecular shape of the compound PH_3 is predicted as being ( )
    - (a) planar
    - (b) linear
    - (c) pyramidal
    - (d) something else

22. Which species is most likely to be planar? ( )
    - (a) NH_4^+
    - (b) CO_3^{2-}
    - (c) SO_3^{2-}
    - (d) ClO_3^-

23. The electron pairs on Cl can be considered as being approximately tetrahedrally directed in ( )
    - (a) ClO^-
    - (b) ClO_2^-
    - (c) ClO_4^-
    - (d) all of these

24. Indicate which one of the following is definitely polar: ( )
    - (a) O_2
    - (b) CO_2
    - (c) BF_3
    - (d) C_2H_3F

25. sp^3 hybridization is important in describing the bonding in ( )
    - (a) H_3O^+
    - (b) CCl_4
    - (c) NH_4^+
    - (d) all of these

26. Although it is known that the actual bonding sequence of atoms in the nitrous oxide molecule is N—N—O, an electronic structure based on the sequence N—O—N (i.e., oxygen bonded to two nitrogens) ( )
    - (a) cannot be drawn
    - (b) can also be drawn, and predicts a polar molecule
    - (c) can also be drawn, and involves unpaired electrons
    - (d) can also be drawn, and obeys the octet rule

27. The hybridization of boron in B_2H_4 is expected to be ( )
    (a) sp
    (b) sp^2
    (c) sp^3
    (d) something else

28. What type of hybrid orbitals is used by carbon in ( ) formaldehyde, CH_2O?
    (a) sp
    (b) sp^2
    (c) sp^3
    (d) none

29. The description of the electronic structure of O_2 that best ( ) accounts for both the bond energy and magnetic properties is given by
    (a) :Ö=Ö:
    (b) :Ö—Ö:
    (c) a resonance hybrid of structures (a) and (b)
    (d) molecular orbital theory

30. All chemical bonds are the result of ( )
    (a) the magnetic interaction of electrons
    (b) the interaction of nuclei
    (c) differences in electronegativity
    (d) the interaction of electrons and nuclei

31. The electrons generally involved in bonding ( )
    (a) are those that lie closest to the nucleus
    (b) are those for which the ionization energies are small
    (c) end up being transferred from one atom to another
    (d) occupy s atomic orbitals

32. The fact that ionization energies for "valence" electrons in ( ) units of kJ/mol (rather than joules per atom) are of the same order of magnitude as heats of reaction for chemical changes (kJ/mol)
    (a) is a mere coincidence involving units
    (b) supports the idea that only the outer-level electrons are generally involved in molecular rearrangements
    (c) supports the idea that chemical changes involve moles of atoms and not simply individual atoms
    (d) has no apparent rationale

33. Which one of the following would have a Lewis structure ( ) most like that of CO_3^{2-}?
    (a) CO_2
    (b) SO_3^{2-}
    (c) NO_3^-
    (d) O_3

34. The weakest N to N bond among the following is most ( ) likely that in
    (a) :N≡N:
    (b) F—N̈=N̈—F

(c) F—N̈—N̈—F
     | |
     F F

(d) H—N̈=N=N̈:

35. A bond angle of 120° appears in species with which ( ) geometry?
  (a) linear
  (b) square
  (c) triangular
  (d) tetrahedral

36. Which of the following is a correct Lewis structure for ( ) C_2H_4O?

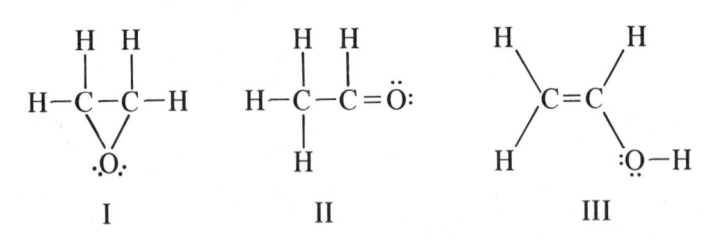

    I   II   III

(a) II
(b) III
(c) II, III
(d) I, II, III

37. Consider the geometry of the carbonate ion, CO_3^{2-}. The ( ) three oxygen atoms lie
  (a) in a straight line
  (b) at three corners of a triangle
  (c) at three corners of a square
  (d) at three corners of a tetrahedron

38. Rubidium carbonate has the formula ( )
  (a) $RbCO_3$
  (b) Rb_2CO_3
  (c) $Rb_2(CO_3)_3$
  (d) $Rb(CO_3)_2$

39. $(NH_4)_2SO_4$ is the formula for ( )
  (a) sulfuric acid
  (b) diammonia sulfate
  (c) sulfate of ammonia
  (d) ammonium sulfate

40. $KClO_4$ is a common ionic substance. The anion has the ( ) formula
  (a) $KClO_4$
  (b) K^+
  (c) Cl^-
  (d) ClO_4^-

41. In which of the following would X show sp^2 hybridization? ( )
  (a) A—X—A
  (b) A=X=A
  (c) A=Ẍ—A
  (d) A—Ẍ—A

42. For which species would the Lewis structure show a double ( )
bond between two carbon atoms?
 (a) C_2H_5F
 (b) C_2H_3F
 (c) C_2HF
 (d) C_2F_2

43. Which of the following resonance forms of CO_2 is the least ( )
plausible and should therefore make the smallest contribution to the
overall structure?
 (a) :Ö—C≡O:
 (b) :O≡C—Ö:
 (c) :Ö=C=O:
 (d) :Ö—C—Ö:

Problems

44. For each of the following species, draw Lewis structures, predict
the hybridization of the central atom, and indicate the molecular geometry
(linear, bent, equilateral triangle, pyramid, tetrahedron).
 (a) SCl_2
 (b) SO_3^{2-}
 (c) NO_2^+

45. Consider the N_2O molecule.
 (a) How many valence electrons does it have?
 (b) Given that the two nitrogen atoms are bonded to each other and that the octet rule is obeyed, draw a Lewis structure for N_2O.
 (c) Describe the geometry of N_2O, based on the structure in (b).
 (d) Would you expect N_2O to be polar?
 (e) Draw a resonance form for N_2O, different from the one shown in (b).

46. Complete the following table:

Compound	Formula	Electron Configuration of Monatomic Ion
Ammonium fluoride	NH_4F	$1s^2\,2s^2\,2p^6$
Calcium hydroxide	_____	_____
Potassium phosphate	_____	_____
_____	$FeSO_4$	_____
Aluminum nitrate	_____	_____

116 • 8–Chemical Bonding

*47. (You will need to consult references other than the text to answer the following.)

What experimental methods would be used to determine such molecular properties as
- (a) the internuclear distance in a gaseous molecule, say NO(g)
- (b) the bond angle in $NO_2(g)$
- (c) the bond energy for $N_2(g)$
- (d) the frequency of rotation of a gaseous diatomic molecule at 298 K, say $N_2(g)$
- (e) the frequency of vibration for the two atoms along the internuclear axis, as in $N_2(g)$ at 298 K

Use this information to help complete the table on the following page.

*48. By reference to molecular orbital diagrams
- (a) arrange O_2^+, O_2, O_2^-, and O_2^{2-} in order of increasing bond energy and decreasing bond length. Consult one or more of the suggested readings to check your predictions against the observed properties.
- (b) interpret the quantity, 1170 kJ/mol, the ionization energy of O_2. (Note that the ionization energy of O is 1310 kJ/mol.)

TABLE 1 SOME EXPERIMENTAL METHODS OF MOLECULAR STRUCTURE DETERMINATION*

I. Diffraction Methods

Radiation absorbed/emitted	Phase studied	Property determined
Electron	gas; solid surface	
Neutron		H atom position
X-ray	solid	

II. Spectroscopic Methods

Radiation absorbed/emitted	Wavelength (m)	Energy (J/photon); (kJ/mol)	Effect on molecule; property determined
X-ray	10^{-8}	; 10^4	;
Ultraviolet (UV)		;	Electronic transition;
Visible (Vis)	$4-7 \times 10^{-7}$	; 10^2	;
Infrared (IR)		;	Vibrational transition;
Microwave; electron paramagnetic resonance (epr)	10^{-2}	;	;
Radio; nuclear magnetic resonance (nmr)	10	; 10^{-5}	Nuclear spin transition;

*See Problem 47 and the Readings at the end of this chpater.

SELF-TEST ANSWERS

1. T
2. F (Note that hybridization often results in the unpairing of electrons: consider Be in BeF_2. Also see Chapter 21.)
3. F ($AlBr_3$, with the expected ions Al^{3+} and Br^-.)
4. T
5. F (Bond length decreases.)
6. T (Consider the atom ratio of B to O needed for octets; assume that ions form.)
7. F (Ionic: combination is of a metal and a nonmetal.)
8. T (Example of reasonable structure: $:\ddot{B}=A=\ddot{B}:$)
9. F (Don't forget to count the two electrons contributed by the charge.)
10. T (See Chapter 7.)
11. b (Electrons are removed first from 4s, not 3d.)
12. d (That is, S is the most electronegative. Review trends in electronegativity, Chapter 7.)
13. c
14. a (An ionic bond holds Na^+ to OH^-; OH^- is held together by a covalent bond.)
15. c (Subtract the first equation from the second; treat ΔH's in the same way, Chapter 4.)
16. d (Odd number of valence electrons.)
17. b (Smallest Δ(EN), Chapters 7 and 8.)
18. a (More than one Lewis structure can be drawn.)
19. d (Perhaps the best representation of the observed properties is $:\ddot{F}-Be-\ddot{F}:$)
20. a (These two hybrid orbitals extend to either side of the central atom, along a straight line.)
21. c (Of four electron pairs on P, one is unshared.)
22. b (Draw Lewis structures and then apply electron pair repulsion ideas.)
23. d (There are four pairs on Cl in each. Geometries would be described as linear, bent, and tetrahedral, respectively.)
24. d (Draw the Lewis structure, predict the geometry, and then predict polarity.)
25. d (Each has four pairs of electrons on the central atom.)
26. d (For example, $:\ddot{N}=O=\ddot{N}:$, which would be linear and nonpolar.)
27. b
28. b (Actual structure,
$$H-\underset{\underset{\|}{\overset{:\ddot{O}:}{}}}{C}-H$$
)

29. d
30. d (See Question 1.)
31. b (So-called valence electrons.)
32. b (One can think of the formation of an ionic bond as involving ionization.)
33. c (They are isoelectronic.)
34. c (See Question 4.)
35. c
36. d (All obey the octet rule. For a discussion of isomers, which this illustrates, see Chapter 10.)
37. b
38. b (The ions expected are Rb^+ and CO_3^{2-}. Apply the principle of electroneutrality.)
39. d (The best name; though one could call it by (c), the cation is normally written first and named first.)
40. d (Become familiar with some of the common ions — their formulas and names.)
41. c (Corresponding to three "pairs" of electrons on X.)
42. b (In each case, start by drawing a carbon-carbon bond, and not things like C—F—C. Another rule based on observation: C frequently bonds to C. More in Chapter 10.)
43. d (Note the lack of an octet on the carbon.)
44. (a) :Cl̈:S̈:Cl̈:, sp^3, bent

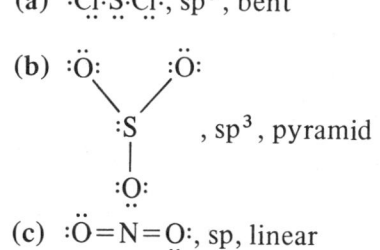

   (b) :Ö: :Ö:
         \\ /
         :S   , sp^3, pyramid
         |
         :Ö:

   (c) :Ö=N=O:, sp, linear

45. (a) 16 (b) :N̈=N=Ö: (c) linear (d) yes (e) :N≡N—Ö:
46. $Ca(OH)_2$, $1s^2 2s^2 2p^6 3s^2 3p^6$
    K_3PO_4, same as Ca^{2+}
    iron(II) sulfate, $[Ar] 3d^6$
    $Al(NO_3)_3$, $1s^2 2s^2 2p^6$
*47. (a) electron diffraction or IR-microwave spectroscopy
     (b) electron diffraction
     (c) x-ray, UV spectroscopy
     (d) microwave spectroscopy
     (e) IR spectroscopy
*48. (a) Bond energy increases (and bond length decreases) in the order: $O_2^{2-} < O_2^- < O_2 < O_2^+$. (The numbers of electron-pair bonds are 1, 1.5, 2, and 2.5, respectively.) Bond lengths range from 0.149 to 0.112 nm.

(b) The electron most easily removed from O_2 is in a π_{2p}^* orbital, higher in energy than the parent 2p orbital from which it is derived.

SELECTED READINGS

Alternative and extensive discussions of bonding, the theories and their history are given in the books by Lewis, Pimentel, and Sisler, listed in Chapter 6, and:

Companion, A. L., *Chemical Bonding*, New York, McGraw-Hill, 1964.
Coulson, C. A., *The Shape and Structure of Molecules*, New York, Oxford, 1973.
Ferreira, R., Molecular Orbital Theory: An Introduction, *Chemistry* (June 1968), pp. 8-15.
Gray, H. B., *Chemical Bonds*, Menlo Park, Ca., W. A. Benjamin, 1973.
Lagowski, J. J., *The Chemical Bond*, Boston, Houghton Mifflin, 1966.
Pauling, L., *The Nature of the Chemical Bond*, Ithaca, N. Y., Cornell University Press, 1960.

As suggested by Self-Test Question 47, our knowledge of molecular structure and dynamics (molecules are constantly active; and energies of rotation and vibration, as well as electronic energies, are quantized) is considerable. Its experimental determination is considered in the discussion of x-ray diffraction in Chapter 11 and in:

Barrow, G. M., *The Structure of Molecules: An Introduction to Molecular Spectroscopy*, New York, W. A. Benjamin, 1963.
Brey, W. S., Jr., *Physical Methods for Determining Molecular Geometry*, New York, Reinhold, 1965.
Sonnessa, A. J., *Introduction to Molecular Spectroscopy*, New York, Reinhold, 1966.
Wagner, J. J., Nuclear Magnetic Resonance Spectroscopy – An Outline, *Chemistry* (March 1970), pp. 13-15.

Octets – just how stable are they? Consider:

Moody, G. J., A Decade of Xenon Chemistry, *Journal of Chemical Education* (October 1974), pp. 628-630.
Selig, H., The Chemistry of the Noble Gases, *Scientific American* (May 1964), pp. 66-77.
Ward, R., Would Mendeleev Have Predicted the Existence of XeF_4?, *Journal of Chemical Education* (May 1963), pp. 277-279.

The shapes of molecules, of molecular orbital contour surfaces (analogous to the orbitals we have drawn for atoms), and the use of models are considered in the article by Wahl listed in Chapter 6 and:

Hall, S. K., Symmetry, *Chemistry* (March 1973), pp. 14-16.
Morrow, F. J., Do-it-yourself Molecular Models, *Chemistry* (April 1972), pp. 6-9.
Pauling, L., *The Architecture of Molecules*, San Francisco, W. H. Freeman, 1964.

9
PHYSICAL PROPERTIES AS RELATED TO STRUCTURE

QUESTIONS TO GUIDE YOUR STUDY

1. How does modern chemical theory explain the properties of bulk samples of matter? What factors at the atomic-molecular level determine these properties?

2. What properties are characteristic of ionic substances? How do these properties vary with the sizes and charges of ions? (What determines the number of electrons an atom gains or loses to form an ion?)

3. What are the distinguishing properties of covalently bonded species? How are they explained? What kind of experimental observation would tell you a substance is covalently bonded?

4. What factors favor the formation of small, discrete molecules; of very large, extensive molecules? Is there a correlation with the Periodic Table?

5. How would you distinguish experimentally between very small and very large molecules? What properties are characteristic of macromolecular and network substances? (What species constitute the structural units in the bulk of the earth's mantle?)

6. How are metallic properties, such as high electrical conductivity, explained in terms of bonding? How strong is the metallic bond?

122 • 9–Physical Properties as Related to Structure

7. What kind of experimental evidence would tell you that you were dealing with interatomic forces; with intermolecular forces? What are the relative magnitudes of the energies required to overcome these forces?

8. Can bulk properties be correlated with position in the Periodic Table, and thus with electron configurations of the constituent atoms?

9. What effects on physical properties are traceable to differences in molecular geometry and molecular polarity?

10. Can you now explain more fully, in terms of interatomic and intermolecular forces, what you directly observe during a chemical reaction? What is happening among the molecules that results in the gross changes you detect?

11.

12.

YOU WILL NEED TO KNOW

Concepts

1. How to write Lewis structures — Chapter 8.
2. How to recognize whether a substance is likely to be molecular, ionic, or metallic (a prediction based on position in the Periodic Table and/or on electronegativities) — Chapters 7, 8.
3. How to predict the geometry and polarity of molecules — Chapter 8.

Math

Another chapter without math!

CHAPTER SUMMARY

In this chapter, we continue the discussion of chemical bonding and extend the atomic-molecular model to include the weaker forces between molecules which account particularly for physical properties.

We recognize three different types of intermolecular attractive forces, whose magnitudes determine properties of molecular substances — such properties as melting point, vapor pressure, and heat of vaporization:

1. *Dipole forces*, which operate between polar molecules and account for the fact that these substances ordinarily have higher melting and boiling points than do nonpolar substances of comparable molecular mass.

2. The *hydrogen bond*, a particularly strong type of dipole force, which arises between molecules in which hydrogen is bonded to a small, highly electronegative atom (N, O, F). These bonds are found in a few inorganic compounds (e.g., H_2O, NH_3, HF, HCN) and a wide variety of organic compounds, including alcohols, acids, and, perhaps most important of all, proteins (Chapters 10, 25).

3. *Dispersion forces,* which operate between all molecules, polar and nonpolar. These forces, arising from the distortion of the electron density of a molecule by its neighbor, increase in magnitude with molecular size. This explains the general observation that melting and boiling points increase with molecular mass.

We discussed the structure and properties of *macromolecular* substances, where large numbers of atoms are held together by covalent bonds to form a huge molecular aggregate. Many familiar substances fall into this category. They include the allotropic forms of carbon, silicon dioxide, and related compounds (quartz, sand, glass, asbestos, and others) and such familiar organic polymers (Chapter 25) as polyethylene and nylon. We have attempted to show in this chapter how the underlying two- or three-dimensional structure of these substances determines their properties and hence their uses.

Metallic bonds are found in elementary substances of low electronegativity (the metals) and in certain alloys of these elements. Metal atoms have too few valence electrons to form covalent bonds with all of their neighbors. Instead, valence electrons are pooled to form a negatively charged "glue" that holds the metal cations together. This simple picture of metallic bonding, which implies high electron mobility, offers a reasonable qualitative explanation of many of the general properties of metals.

BASIC SKILLS

1. **Compare the melting points of different ionic compounds.**

See Table 9.1 in the text and the discussion immediately preceding it. In general, the highest melting ionic compounds are those containing small, highly charged, monatomic ions. Note that compounds containing polyatomic ions (notably OH^-, CO_3^{2-}) frequently decompose before they melt.

This skill is required to work Problems 9.1 and 9.21.

124 • 9–Physical Properties as Related to Structure

   2. **Compare the melting points or boiling points of different molecular substances.**

The principles involved here are:

   a. among substances of similar structure, melting point or boiling point ordinarily increases with molecular mass;
   b. among substances of comparable molecular mass, those which are more polar or less compact ordinarily boil higher;
   c. hydrogen bonded species have unusually high boiling points.

Example 9.1 illustrates the first two of these principles. See Problems 9.2, 9.9, and 9.21.

   3. **Classify molecular substances as to the types of intermolecular forces present.**

There are three types of intermolecular forces:

   a. dispersion forces, common to all molecular substances;
   b. dipole forces, restricted to polar molecules;
   c. hydrogen bonds, found in polar molecules where H is bonded to F, O, or N.

See Example 9.2. Note that in order to use this skill, you must first decide whether the molecule in question is polar or nonpolar. Try Problems 9.3 and, in part, Problems 9.12 and 9.24 (several of the substances listed in these problems are not molecular).

   4. **Classify a substance as to the type of bonding present and make qualitative predictions concerning its physical properties.**

This skill is basic to most of the problems at the end of Chapter 9. In its simplest form, it is used in Problems 9.4, 9.8, 9.15, 9.20, and 9.27. Perhaps less obviously, it is required to work Problems 9.10, 9.11, 9.12, 9.22, 9.23, and 9.24. Before you attempt to work these problems, you must decide whether the substances involved are ionic, molecular, macromolecular, or metallic.

The basic principles involved in classifying substances as to bond type are covered in Table 9.4 of the text. This table, repeated here, is well worth your careful study and understanding; it summarizes much of the material presented earlier in the chapter.

In addition to the four skills listed above, a considerable amount of descriptive chemistry is covered in this chapter. You should be familiar with:

TABLE 2 – MOLECULAR STRUCTURE AND PROPERTIES OF BULK MATTER

Structural Units	Forces Within Units	Forces Between Units	Properties	Examples
1. Ions	Covalent bond in polyatomic ions (e.g., CO_3^{2-})	Ionic bond	Hard, brittle solid (high mp, bp); conducts in melt and water solution; soluble in polar solvent	$NaCl$, MgF_2, $CaCO_3$
2. Molecules (a) nonpolar	Covalent bond	Dispersion force	Soft solid, or liquid or gas (low mp, bp); nonconductor; soluble in nonpolar solvent	H_2, CH_4, BF_3
(b) polar	Covalent bond	Dispersion force and dipole force; hydrogen bond	Low mp, bp (but higher than for nonpolar); often conducts in water solution; more soluble in polar solvent than nonpolar	HCl, NH_3, CH_3OH
3. Macromolecule, network	Covalent bond; often, any other as well	... Dispersion force, dipole force, hydrogen bond may be present	Hard solid (high mp, bp); may decompose on heating; usually nonconductor; insoluble	C, SiO_2, cellulose
4. Cations and mobile electrons	...	Metallic bond	Variable mp, high bp, variable hardness; good conductor; insoluble except in each other	Na, Fe, brass

Approximate order of bond strength: Dipole < Dispersion < Hydrogen ≪ { Ionic, Covalent, Metallic }

- the decomposition reactions observed when certain ionic compounds are heated (Section 9.1; Problems 9.13 and 9.25).
- the structures of certain important macromolecular substances (diamond, graphite, silica, and the silicates; see Problems 9.14 and 9.26).
- the general properties of metals (Section 9.4).

SELF-TEST

True or False

1. NaOH(s) is a poor electrical conductor because it is made up ( ) of neutral molecules.

2. The physical properties of molecular substances are directly ( ) related to the strengths of the covalent bonds holding together the molecules.

3. Boiling points of molecular substances usually increase with ( ) molecular mass.

4. Forces between atoms increase in the order dipole < ( ) hydrogen bond < ionic bond.

5. Diamond and graphite are *isomeric* forms of carbon. ( )

6. An increase in pressure increases the stability of diamond ( ) relative to graphite.

7. Carbonates and other substances containing polyatomic ions ( ) frequently decompose on heating.

8. High charge density (small size, high charge) leads to strong ( ) interionic forces of attraction; low charge density, weak forces.

9. Low solubility in common solvents is characteristic of polar ( ) molecules.

10. High malleability is a distinguishing property of pure metals. ( )

Multiple Choice

11. Which one of the following solids consists of small, discrete ( ) molecules?
    (a) graphite   (b) dry ice
    (c) iron   (d) NaCl

12. Of the following interactions at the atomic-molecular level, ( ) the strongest is most likely to be

(a) dispersion force (b) dipole force
(c) hydrogen bond (d) metallic bond

13. Which one of the following substances could be boiled ( )
without breaking hydrogen bonds?
    (a) H_2 (b) NH_3
    (c) H_2O_2 (d) none of these

14. Which kind of force is *not* overcome in boiling a sample of ( )
water?
    (a) hydrogen bond (b) dispersion force
    (c) dipole force (d) something else

15. Molecules themselves are broken apart, during at least some ( )
phase changes, in the case of
    (a) SiC (b) sulfur
    (c) graphite (d) all of these

16. In which pair of substances must the same type of attractive ( )
force be overcome when
    (a) boiling H_2 and HF
    (b) melting C and Ca
    (c) dissolving LiCl and ICl
    (d) melting CCl_4 and I_2

17. Which of the following would have a boiling point lower ( )
than that of $SiCl_4$?
    (a) $GeCl_4$ (b) $SiBr_4$
    (c) CCl_4 (d) LiCl

18. Which of the following is a gas at room temperature? (The ( )
others are liquid or solid at room temperature.)
    (a) C_4H_{10} (b) C_5H_{12}
    (c) C_3H_7OH (d) $AlCl_3$

19. Metallic substances are distinguished from other classes of ( )
chemical substances by their
    (a) high boiling points
    (b) being solids at 25°C
    (c) high electrical conductivity in the liquid state
    (d) high electrical conductivity in the solid state

20. Which should be the best electrical conductor? ( )
    (a) graphite (b) silver
    (c) ice (d) quartz

21. Two-dimensional, or sheet-like, structures are observed in ( )
    (a) graphite (b) diamond
    (c) ice (d) all of these

22. Ammonia, NH_3, is more soluble in water than is methane, ( )
CH_4. This is most likely a result of
    (a) molecular mass difference
    (b) difference in melting points
    (c) a difference in densities
    (d) hydrogen bonding

23. The various types of glass you use in the laboratory are ( )
unlikely to contain appreciable amounts of
    (a) boron   (b) hydrogen
    (c) oxygen   (d) sodium

24. Color in a substance is often associated with ( )
    (a) unpaired electrons
    (b) alternating single and double bonds
    (c) compounds of transition metals
    (d) all of these

25. The fact that NH_3 boils at a higher temperature than does ( )
PH_3 can be explained by noting that NH_3
    (a) has a smaller molecular volume
    (b) exhibits dipole forces
    (c) has larger bond angles
    (d) exhibits hydrogen bonding

26. A certain substance has a low mp, is slightly soluble in water ( )
but more soluble in CCl_4, and does not conduct electricity. It is most
likely
    (a) ionic   (b) molecular
    (c) macromolecular   (d) metallic

27. Diamond and quartz are similar in that both ( )
    (a) are allotropes of carbon
    (b) are good conductors
    (c) carbon and silicon are tetrahedrally bonded to four other atoms
    (d) are synthetic polymers

28. A given substance is a blue solid at 25°C. It decomposes at a ( )
temperature below 300°C to give a white solid and a second product
which is volatile. The blue solid dissolves in water to give a
conducting solution. If the blue solid is known to be one of the
following, it is most likely
    (a) Cu   (b) $CuSO_4 \cdot 5\ H_2O$
    (c) $CuCl_2$ (d) an alloy of Ag and Cu

29. Which substance would you expect to have the highest ( )
boiling point?
    (a) Cl_2 (b) BrCl
    (c) HCl   (d) NaCl

30. Which factors tend to favor high melting point in an ionic ( ) solid?
   (a) small ionic radius
   (b) high ionic charge
   (c) high formula mass
   (d) high charge density

31. A water solution of $MgSO_4$ might be expected to contain ( ) the species
   (a) Mg^+ (b) SO_3
   (c) SO_4^{2-} (d) any of these

32. The sharing of three of the oxygens in each SiO_4 tetra- ( ) hedron leads to structures which are
   (a) single-stranded chains
   (b) double-stranded chains
   (c) planar sheets
   (d) three-dimensional networks

33. Silicates may exhibit ( )
   (a) covalent bonding   (b) ionic bonding
   (c) dispersion forces   (d) all of these

34. Magnesium carbonate, $MgCO_3$, decomposes on heating. The ( ) reaction is
   (a) $MgCO_3(s) \rightarrow MgO(s) + CO_2(g)$
   (b) $MgCO_3(s) \rightarrow Mg(s) + CO_2(g) + \frac{1}{2}O_2(g)$
   (c) $MgCO_3(s) \rightarrow MgO_2(s) + CO(g)$
   (d) $MgCO_3(s) \rightarrow Mg(s) + CO(g) + O_2(g)$

Problems

35. Classify each of the following substances as likely to be ionic, molecular, macromolecular, or metallic.

    A is a high-melting solid (mp = 1100°C); a 50-g sample dissolves completely in 100 g water_____

    B is a high-melting solid, insoluble in water, and an excellent conductor of heat_____

    C is a gas at 25°C and 1 atm_____

    D is a solid, soluble in benzene but not in water_____

    E melts at 2400°C; neither the solid nor the melt conducts electricity_____

130 • 9–Physical Properties as Related to Structure

36. Consider the following types of forces: ionic bonds, covalent bonds, hydrogen bonds, dipole forces, and dispersion forces. Which of these would have to be overcome to melt

(a) $MgCl_2$
(b) CCl_4
(c) SiO_2
(d) HF
(e) KOH

*37. Prepare a discussion on: "The relation of odor to molecular geometry."

*38. The band theory of metals is a more sophisticated treatment of metallic bonding. How does band theory account for such characteristic properties as high thermal conductivity, luster, and malleability?

SELF-TEST ANSWERS

1. F (Ions are more or less fixed in position at ordinary temperatures.)
2. F (Intermolecular forces are determining factors.)
3. T (A result of larger dispersion forces.)
4. T
5. F (Allotropic forms.)
6. T (The more compact arrangement of atoms is favored at high pressure.)
7. T
8. T (The coulombic force of attraction is directly proportional to the charges, inversely proportional to their separation.)
9. F (Macromolecular and metallic substances are usually insoluble.)
10. T (Another commonly known property of metals.)
11. b (Dry ice is solid CO_2.)
12. d (Strongest by an order of magnitude.)
13. a (For hydrogen bonding to exist, H must be bonded to a highly electronegative atom.)
14. d (Specifically, a polar covalent bond. *Note:* molecules generally remain intact during phase changes; weaker, intermolecular forces are overcome before chemical bonds.)
15. d (SiC and graphite are macromolecular; sulfur happens to contain cyclic S_8 molecules which break open easily.)
16. d (Dispersion forces only.)
17. c (Lower molecular mass.)
18. a (Nonpolar, low molecular mass.)
19. d (This property is common in no other class.)
20. b (A metal.)
21. a (The others have three-dimensional networks.)

22. d (In both NH_3 and H_2O. More on solubility principles in Chapter 12.)
23. b
24. d (Examples for (a) are O_2, NO_2; (b) most organic dyes; (c) – see Chapters 7, 21.)
25. d
26. b (Low mp rules out ionic; solubility rules out macromolecular and metallic.)
27. c
28. b (The volatile product is the water of hydration.)
29. d (Ionic.)
30. d (See the answer to Question 8 above.)
31. c (Also Mg^{2+} and H_2O. How should you know that magnesium will give a +2 ion? See Chapter 8.)
32. c (As in talc and mica.)
33. d (Examples: covalent bond, Si—O; ionic bond, cations plus anionic chain or sheet of SiO_4 tetrahedra; dispersion force, between chains or sheets.)
34. a (Analogous to $CaCO_3$.)
35. A is ionic; B, metallic; C, molecular; D, molecular; E macromolecular.
36. (a) ionic bonds   (b) dispersion forces
    (c) covalent bonds   (d) hydrogen bonds and dispersion forces
    (e) ionic bonds
*37. To begin, try reading the article by Amoore, et al., noted below. Key words to be used in locating the literature articles would include *odor*, *stereochemical theory*, and *pheromone*.
*38. References might include a textbook of physical chemistry; Holden, A., *The Nature of Solids*, listed below; and other sources suggested by your instructor. First review the discussion of molecular orbital theory in Chapter 8.

SELECTED READINGS

Another discussion of intermolecular forces is given in:

House, J. E., Jr., Weak Intermolecular Interactions, *Chemistry* (April 1972), pp. 13–15.

On the properties of diamond and other materials:

Bundy, F. P., Superhard Materials, *Scientific American* (August 1974), pp. 62–70.
Derjaguin, B. V., The Synthesis of Diamond at Low Pressure, *Scientific American* (November 1975), pp. 102–109.
Materials, *Scientific American* (September 1967).

Shapes of molecules and their odors:

Amoore, J. E., The Stereochemical Theory of Odor, *Scientific American* (February 1964), pp. 42–49.

Schneider, D., The Sex-Attractant Receptor of Moths, *Scientific American* (July 1974), pp. 28–35.

Metallic bonding as well as other solid-state properties (see Chapter 11 of the text) are discussed in:

Holden, A., *The Nature of Solids*, New York, Columbia University Press, 1965.

Moore, W. J., *Seven Solid States: An Introduction to the Chemistry and Physics of Solids*, New York, W. A. Benjamin, 1967.

10
AN INTRODUCTION TO ORGANIC CHEMISTRY

QUESTIONS TO GUIDE YOUR STUDY

1. What properties are unique to organic compounds?
2. What elements are found in organic compounds? What kinds of interatomic and intermolecular forces are present in these substances?
3. How do you account for the vast number and variety of carbon compounds? How is carbon different from the other Group 4A elements?
4. How is it possible for two or more different compounds to have the same molecular formula?
5. How do isomers differ from one another? Are these differences predictable in terms of molecular structure?
6. How does the electronic structure of a bonded group of atoms, such as $>C=O$, determine the physical and chemical properties characteristic of that group?
7. Having classified many organic compounds according to these functional groups, what are the reactions characteristic of each class? Besides combustion, what other common reactions of organic materials can you think of?
8. What are the natural and synthetic sources of each class of organic compound? Are the natural sources renewable?
9. What synthetic products can you list which can be labelled as organic?
10. How do soaps and detergents work? What are their structural differences? What are their relative advantages and disadvantages?

11.

12.

YOU WILL NEED TO KNOW
Concepts

1. How to write and interpret Lewis structures; how to predict molecular geometry — Chapter 8.
2. How to relate physical properties to intermolecular forces — Chapter 9.

Math

To work some of the problems of this chapter :
1. How to estimate ΔH from bond energies — Chapter 4.
2. How to make stoichiometric calculations, from formulas and equations — Chapter 3.

CHAPTER SUMMARY

This chapter offers a change of pace, turning from the consideration of theoretical principles (especially those of Chapters 4, 8, and 9) to their illustration in the structures, physical and chemical properties, sources, and uses of organic compounds. Organic compounds are of major interest and importance because of their dominant role in the chemistry of life; the innumerable uses found for them in chemotherapy, as dyes, cleansing agents, perfumes, clothing, fuel, etc. — the vast amounts of materials that support a vigorous chemical technology. (Some 10^{10} kg of ethylene and about 10^7 kg of aspirin were produced in the U. S. in 1975.)

The properties of many organic compounds have been observed and studied for centuries: fermentation of sugar to yield alcohol; air oxidation of alcohol to give sour acetic acid; the preparation of salts of fatty acids (soaps); and the isolation of indigo blue dye — these are age-old practices in organic chemistry.

The simplest organic compounds are the hydrocarbons, which contain only hydrogen and carbon. Among them, we distinguish between paraffins or alkanes, in which all the bonds between carbon atoms are single (e.g.,

CH_4, C_2H_6, . . . , C_nH_{2n+2}); olefins or alkenes, where there is a double bond (e.g., C_2H_4, C_3H_6, . . . , C_nH_{2n}); and acetylenes or alkynes, with a triple bond between two carbons (C_2H_2, C_3H_4, . . . , C_nH_{2n-2}). Still another class of hydrocarbons is made up of the aromatics, which may be considered as derivatives of benzene, C_6H_6.

Many of the most familiar organic substances contain the three elements carbon, hydrogen, and oxygen. Such compounds are conveniently classified according to the functional group present. For example, all alcohols contain the group —C—O—H, and all organic acids, the group $-\underset{\underset{O}{\|}}{C}-O-H$. All compounds containing a given functional group show similar physical and chemical properties — within a series of alcohols, or acids, etc., there is a graduation in physical properties with increasing molecular mass; a typical alcohol, R—OH, will react with a typical acid, R'—COOH, to give an ester, R—O—CO—R'. (Note that the ester functional group, entirely covalently bonded and lacking the ability to participate in hydrogen bonding, imparts its own set of characteristic properties, very different from either parent reactant.) So, classification by functional group allows one to organize a vast amount of organic chemistry.

We have noted the trend toward greater reactivity, from paraffins to olefins to acetylenes. We can see a rationale for this trend when we realize the potential for the formation of additional bonds to carbon when there are double or triple bonds present. (Show that such a reaction is exothermic: $C_2H_4(g) + H_2(g) \rightarrow C_2H_6(g)$.)

One reason for the multiplicity of organic compounds (there are more than 10^6 known!) is the phenomenon of isomerism. We ordinarily find that for a given molecular formula there are several different compounds with quite different properties. Examples include n-butane and the branched iso-butane (molecular formula C_4H_{10}); n-propyl and isopropyl alcohols (C_3H_8O); butyl alcohol and diethyl ether ($C_4H_{10}O$). Noting that each of these compounds contains at least two carbon atoms bonded to each other, we see one reason why isomerism is more common here than in inorganic compounds. Atoms of elements other than carbon very rarely bond *strongly* to one another to give structures of the type —X—X—.

BASIC SKILLS

1. Given the molecular formula of a simple organic compound, draw structural formulas for its different isomers.

This skill is illustrated by Example 10.2 and required to work Problems 10.1, 10.6, 10.7, 10.26, and 10.27. It is best to approach problems such as these in a systematic way. In Problem 10.6, for example, you might start by

136 • 10–An Introduction to Organic Chemistry

drawing the isomer with a 7-carbon chain, then all the isomers with 6-carbon chains, and so on. The most common mistake is to draw too many isomers; you must realize, for instance, that the two members of each of the following pairs are actually identical:

$$\text{C-C-C-C-C-C-C} \quad \text{and} \quad \begin{array}{c} \text{C-C-C-C-C-C} \\ | \\ \text{C} \end{array}$$

(both have a 7-carbon chain)

$$\begin{array}{c} \text{C-C-C-C-C-C} \\ | \\ \text{C} \end{array} \quad \text{and} \quad \begin{array}{c} \text{C-C-C-C-C-C} \\ | \\ \text{C} \end{array}$$

(both have a 1-carbon branch on the second atom in the chain)

A molecular model kit is very helpful in constructing isomers. If you make models of the above pairs, it will be immediately obvious that they are identical.

2. Draw *cis* and *trans* isomers for compounds containing a double bond.

See Example 10.4 and Problems 10.8 and 10.28. Note that in order to have *cis-trans* isomerism, the two groups attached to each of the double-bonded carbon atoms must differ from one another. Thus the compound

$$\begin{array}{c} \text{H} \quad \text{H} \\ | \quad | \\ \text{Cl-C=C-Cl} \end{array}$$

can have a *cis* and a *trans* isomer

$$\begin{array}{c} \text{H} \\ \diagdown \\ \text{Cl} \end{array} \text{C=C} \begin{array}{c} \text{H} \\ \diagup \\ \text{Cl} \end{array} \quad \text{and} \quad \begin{array}{c} \text{H} \\ \diagdown \\ \text{Cl} \end{array} \text{C=C} \begin{array}{c} \text{Cl} \\ \diagup \\ \text{H} \end{array}$$

cis   trans

but the compounds

$$\begin{array}{ccc} \text{H H} & \text{Cl H} & \text{H Cl} \\ | \; | & | \; | & | \; | \\ \text{Cl}-\text{C}=\text{C}-\text{H}, & \text{Cl}-\text{C}=\text{C}-\text{H}, \text{ and } & \text{Cl}-\text{C}=\text{C}-\text{Cl} \end{array}$$

cannot show this type of isomerism.

3. Given the molecular formula of a straight-chain hydrocarbon, classify it as an alkane (paraffin), alkene (olefin), or alkyne (acetylene).

Perhaps the simplest way to make a decision of this type is to look at the number of hydrogen relative to carbon atoms. Thus, we have the general formulas

alkane (all single bonds): $C_n H_{2n+2}$

alkene (one double bond): $C_n H_{2n}$

alkyne (one triple bond): $C_n H_{2n-2}$

Using these relations, you should have no difficulty with Problems 10.9 and 10.29.

4. Draw structural formulas for aromatic organic compounds.

See Example 10.5 and Problem 10.10. Note that in the benzene ring, if we take position 1 as our reference point:

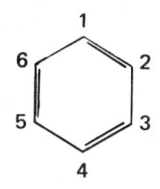

- points 2 and 6 are closest to 1 (two *ortho* positions).
- points 3 and 5 are farther away from 1 (two *meta* positions).
- point 4 is farthest removed from 1 (*para* position).

5. Draw structural formulas for alcohols, ethers, aldehydes, ketones, acids, or esters having a given number of carbon atoms.

See Example 10.6 and Problems 10.17, 10.18, 10.37, and 10.38. Notice that this skill requires that you

10–An Introduction to Organic Chemistry

a. recognize the functional group characteristic of each of these types of oxygen-containing compounds; i.e.,

$$\text{alcohol} \quad -OH \qquad \text{aldehyde} \quad -\underset{|}{\overset{H}{C}}=O \qquad \text{acid} \quad -\overset{O}{\underset{||}{C}}-OH$$

$$\text{ether} \quad -O- \qquad \text{ketone} \quad -\underset{||}{\overset{}{C}}- \qquad \text{ester} \quad -\underset{||}{\overset{}{C}}-OR$$
$$\qquad\qquad\qquad\qquad\qquad\qquad O \qquad\qquad\qquad O$$

b. apply Skill 1 to obtain all possible isomers of a given type of compound.

6. **Given the molecular formula of an oxygen-containing organic compound, classify it as an alcohol, ether, aldehyde, ketone, acid, or ester.**

This skill is illustrated in Example 10.7; note that the type formulas given in Table 10.3 are particularly helpful in making a classification of this sort. Remember also that:
— every alcohol (except CH_3OH) is isomeric with at least one ether.
— every aldehyde (except HCHO and H_3C-CHO) is isomeric with at least one ketone.
— every acid (except $H-COOH$) is isomeric with at least one ester.
See Problems 10.20 and 10.40.

In addition to these specific skills, a considerable amount of descriptive organic chemistry is covered in this chapter. In studying this material you should acquire a general knowledge of:
— the types of chemical reactivity shown by the different classes of hydrocarbons.
— the processes used to obtain gasoline from petroleum.
— the processes under study to obtain low molecular mass hydrocarbons from coal.
— methods used to prepare common organic compounds including acetylene, ethyl alcohol, methyl alcohol, acetaldehyde, acetone, acetic acid, and ethyl acetate.
— the characteristic structures and methods of preparation of soaps and detergents.
— methods used to obtain alkyl halides from hydrocarbons.

SELF-TEST

True or False

1. The melting point of aspirin is about 135°C. You prepare a ( )
sample of the compound in the laboratory and find its melting point
to be about 120°C. You should consider your sample to be impure.

2. The general formula for a cycloparaffin is C_nH_{2n+2}. ( )

3. The molecular formula C_5H_{10} may represent an olefin. ( )

4. A saturated hydrocarbon is one in which each carbon atom ( )
is bonded to four other atoms.

5. The hydrocarbon C_4H_{10} shows *cis-trans* isomerism. ( )

6. Carbon always bonds to four other atoms. ( )

7. The double bond in benzene and other aromatics behaves ( )
chemically in the same way as that in ethylene.

8. Ethyl alcohol, C_2H_5OH, would be expected to show ( )
hydrogen bonding.

9. The gasification of coal is a new process that promises relief ( )
from natural gas shortages.

10. The origins of petroleum are thought to lie in the ( )
decomposition of marine organisms.

Multiple Choice

11. Carbon compounds are so numerous and varied because ( )
    (a) C forms strong bonds with a variety of nonmetal atoms
    (b) C forms strong single as well as stronger multiple bonds
        with itself
    (c) C atoms may bond in rings as well as in chains,
        branched and unbranched
    (d) all the above

12. Choose the class of hydrocarbons that participates in the ( )
largest variety of reactions:
    (a) alkanes   (b) alkenes
    (c) alkynes   (d) olefins

13. A carbon-carbon double bond ( )
    (a) fixes the relative positions of the four atoms bonded to
        the two carbons

140 • 10–An Introduction to Organic Chemistry

    (b) is twice as strong as a single bond
    (c) is half as long as a single bond
    (d) is less reactive than a single bond

14. The geometry of the four atoms in an acetylene, ( )
$-X-C\equiv C-Y-$, is
    (a) linear   (b) bent
    (c) square   (d) tetrahedral

15. An aromatic compound is characterized by   ( )
    (a) an odor   (b) a high melting point
    (c) a benzene ring   (d) a double bond

16. Mere separation of petroleum into its components relies ( )
mostly on the process of
    (a) fractional crystallization
    (b) fractional distillation
    (c) combustion
    (d) cracking

17. The forces overcome in boiling acetic acid, CH_3COOH, ( )
include
    (a) dispersion forces   (b) dipole forces
    (c) hydrogen bonds   (d) all of these

18. Which of the following is likely to have the lowest melting ( )
point?
    (a) C_2H_5OH (b) CH_3COOH
    (c) $CH_3COOC_2H_5$ (d) CCl_3COOH

19. Of the substances represented by the molecular formula ( )
C_3H_6O, the one with the highest boiling point is a(n)
    (a) acid   (b) alcohol
    (c) aldehyde   (d) ketone

20. The total number of isomeric alcohols with the formula ( )
C_4H_9OH is
    (a) two   (b) three
    (c) four   (d) more than four

21. Hydrogen bonding has an important effect on the physical ( )
properties of
    (a) acids   (b) alkanes
    (c) esters   (d) ketones

22. Which one of the following is an ester?   ( )
    (a) $C_2H_5-O-C_2H_5$ (b) $CH_3COC_2H_5$
    (c) CH_3CH_2COOH (d) $C_3H_7COOC_2H_5$

23. Which one of the following should occur as *cis-trans* ( )
isomers?
    (a) C_2H_4 (b) $C_2H_4Cl_2$
    (c) $C_2H_2Cl_2$ (d) $C_6H_4Cl_2$

24. How many different compounds have the molecular ( )
formula C_5H_{12}?
    (a) one   (b) two
    (c) three   (d) some other number

25. Which one of the following should have the highest boiling ( )
point?
    (a) n-C_8H_{18} (b) iso-C_8H_{18}

    (c) cyclooctane, C_8H_{16} (d) $H_3C-\underset{\underset{H_3C}{|}}{\overset{\overset{H_3C}{|}}{C}}-\underset{\underset{CH_3}{|}}{\overset{\overset{CH_3}{|}}{C}}-CH_3$

26. Maple syrup has the property of being syrupy (viscous) as a ( )
result of, at least in part,
    (a) being mostly water
    (b) being more dense than water
    (c) containing lots of macromolecules
    (d) containing lots of alcohol functional groups, R—OH

27. Oxidation of an aldehyde may yield a(n) ( )
    (a) alcohol   (b) ketone
    (c) acid   (d) ether

28. Large numbers and varieties of consumer products are ( )
derived from substances found in
    (a) air   (b) natural gas
    (c) graphite   (d) petroleum

29. The cracking of a mixture of alkanes may result in the ( )
formation of
    (a) H_2 (b) odd-electron molecules
    (c) higher MM alkanes   (d) synthesis gas

30. Synthesis gas has potential value as raw material for the ( )
production of hydrocarbons primarily because it contains large
amounts of
    (a) CH_4 (b) CO_2
    (c) H_2 and CO   (d) synthetic gasoline

Problems

31. A certain organic substance has the molecular formula C_3H_6O. It could be an aldehyde, ketone, or unsaturated alcohol. Draw a structure for each of these isomers and indicate the hybridization of each carbon atom in the isomer.

32. Draw structural formulas for all the isomers of C_4H_8 (no rings).

33. Draw structures for three isomers of C_3H_5COOH, all of which are organic acids.

*34. Both benzene and paradichlorobenzene, [structure: benzene ring with Cl at para positions], are nonpolar. Yet, hydroquinone, [structure: benzene ring with OH and HO at para positions], is polar. Explain these differences and predict the relative melting points and water solubilities of these compounds.

*35. What contribution did Louis Pasteur make to the study of the molecular geometry of organic compounds?

SELF-TEST ANSWERS

1. T (Melting point is a characteristic property; it is lowered by impurities – Chapters 1 and 12.)
2. F (Paraffin or alkane; cycloparaffin would contain two less H atoms.)
3. T (Draw a Lewis structure and see that a double bond is required – Chapter 8.)
4. T
5. F (No double bond is present in the paraffin.)
6. F (Not when there are multiple bonds.)
7. F (Resonance imparts special properties to the benzene ring.)
8. T (H is bonded to the highly electronegative O – Chapter 9.)
9. F (It has been around for most of the century!)
10. T
11. d
12. c (Triple bond most reactive.)
13. a (And all six atoms are coplanar. For predicting geometry, review Chapter 8.)
14. a (Again, this can be predicted from electron pair repulsion – Chapter 8.)
15. c
16. b (The nature of the process is described in Chapter 1.)

17. d
18. c (Hydrogen bonding is possible in the others.)
19. b (The formula cannot represent an acid. Of the others, only the alcohol can be hydrogen bonded. See Problem 31 below.)
20. c (Can you draw all four?)
21. a (The only one with O—H bonding.)
22. d (Which is an ether? A ketone? An acid?)
23. c (The only one with a double bond and the possibility of two kinds of atoms bonded to each carbon.)
24. c (The formula tells you it is saturated; start by drawing chains.)
25. a (The longest molecule.)
26. d (In the sugars present. A guess at this point.)
27. c (Literally add oxygen to the functional group.)
28. d (Especially in this age of petroleum-based plastics.)
29. a (As well as lower MM alkanes, alkenes, and isomers.)
30. c (See the reactions discussed in the chapter.)
31. aldehyde:

$$H-\underset{\underset{H}{|}}{\overset{\overset{H}{|}}{C}}-\underset{\underset{H}{|}}{\overset{\overset{H}{|}}{C}}-\overset{\overset{H}{|}}{C}=O; \; sp^3, sp^3, sp^2$$

ketone:

$$H-\underset{\underset{H}{|}}{\overset{\overset{H}{|}}{C}}-\underset{\underset{O}{\|}}{C}-\underset{\underset{H}{|}}{\overset{\overset{H}{|}}{C}}-H; \; sp^3, sp^2, sp^3$$

alcohol:

$$H-\overset{\overset{H}{|}}{C}=\overset{\overset{H}{|}}{C}-\underset{\underset{H}{|}}{\overset{\overset{H}{|}}{C}}-O-H; \; sp^2, sp^2, sp^3$$

32. $H_2C=\underset{\underset{H}{|}}{C}-CH_2-CH_3$, $H_2C=\underset{\underset{CH_3}{|}}{C}-CH_3$, $\underset{H}{\overset{H_3C}{\diagdown}}C=C\underset{H}{\overset{CH_3}{\diagup}}$,

$\underset{H}{\overset{H_3C}{\diagdown}}C=C\underset{CH_3}{\overset{H}{\diagup}}$

33. $H_2C=\underset{\underset{H}{|}}{C}-CH_2-\underset{\underset{O}{\|}}{C}-OH$, $H_3C-\underset{\underset{H}{|}}{C}=\underset{\underset{H}{|}}{C}-\underset{\underset{O}{\|}}{C}-O-H$ (cis and trans)

*34. Consider the bonding geometry about the group C—O—H. Would you expect these three atoms to be coplanar with the benzene ring? Hydroquinone can exhibit hydrogen bonding and should be the most soluble in water as well as the highest melting.

*35. Optical isomerism.

SELECTED READINGS

The history of organic chemistry is surveyed in:

Benfey, O. T., *From Vital Force to Structural Formulas*, Boston, Houghton Mifflin, 1964.

The scope of organic chemistry, from practical application to the origin of life:

Bailey, M. E., The Chemistry of Coal and Its Constituents, *Journal of Chemical Education* (July 1974), pp. 446–448.

Blumer, M., Polycyclic Aromatic Compounds in Nature, *Scientific American* (March 1976), pp. 35–45.

Clapp, L. B., *The Chemistry of the OH Group*, Englewood Cliffs, N. J., Prentice Hall, 1967.

Dence, J. B., Covalent Carbon-Metal (loid) Compounds, *Chemistry* (January 1973), pp. 6–13.

Miller, S. L., *The Origins of Life on Earth*, Englewood Cliffs, N. J., Prentice-Hall, 1974.

11
LIQUIDS AND SOLIDS; PHASE CHANGES

QUESTIONS TO GUIDE YOUR STUDY

1. What properties do you generally associate with solids, with liquids, and with gases? Can you qualitatively account for the differences?

2. What kinds of substances normally exist as liquids (or solids or gases) at 25°C and one atmosphere? What correlation have you seen with position in the Periodic Table, with molecular geometry and with intermolecular forces?

3. Under what conditions may a given substance exist as a solid or as a liquid? What conditions favor the conversion from one phase to another?

4. Are there simple laws, as there are for gases, relating volume, temperature, and pressure for a liquid or a solid?

5. How would you describe the process of melting — first in terms of what you would observe, then in terms of atomic-molecular behavior? What energy effects accompany these processes?

6. What kind of experiment would you do to show the existence of a vapor above a liquid? Or to show how the pressure of this vapor changes with temperature?

7. How would you recognize boiling? What would you measure?

8. How do we know the arrangements of atoms and molecules in solids and liquids? How do we account for the regularities of shape and size of crystals?

9. How do you account for the fact that some solids, like dry ice, are converted directly to gases, without first melting?

10. What properties of solids and liquids are important to the activities of an architect, a ham radio operator, or a goldfish in a frozen pond?

11.

12.

YOU WILL NEED TO KNOW
Concepts

1. The meaning of ΔH and the interpretation of its sign; how to use Hess's law — Chapter 4.
2. The ideas of kinetic theory about the nature of a gas (this chapter extends kinetic theory to liquids and solids) — Chapter 5.
3. The kinds of intermolecular forces expected for any given substance (which assumes you know how to predict molecular geometry, electronic structure . . .) — Chapters 6, 8, and 9.

Math

1. How to work with linear equations and their graphs — see the Preface to this guide.
2. How to find a logarithm and an antilogarithm — see the text, Appendix 4.
3. How to use the Ideal Gas Law — Chapter 5.
4. The geometry of a cube: how to relate (e.g., by using the Pythagorean theorem) the dimensions — edge, body and face diagonals (for a cube of edge ℓ, face diagonal = $\sqrt{2}\ell$, body diagonal = $\sqrt{3}\ell$).

CHAPTER SUMMARY

The principal distinction between the gaseous state of matter, discussed in Chapter 5, and the condensed states (liquid and solid) considered in this chapter is the distance of separation between molecules. In the gas state, at ordinary temperatures and pressures, the molecules themselves account for a negligible fraction of the total volume. In the condensed states, the molecules ordinarily occupy from 50 to 70 per cent of the total volume. This structural difference explains why liquids and solids have much greater densities than gases, why they are less compressible, and expand less on heating.

Since molecules in the condensed states are closer together, short-range intermolecular forces exert a much greater influence on physical behavior. We have seen (Chapter 9) how the strength of these forces depends upon the type of molecule present. This is why we cannot write a simple equation of state, analogous to the Ideal Gas Law, to describe the physical behavior of all liquids or all solids. Each liquid and solid has a characteristic density, compressibility, and expansibility.

The major distinction between the liquid and solid states is one of molecular mobility. The particles in a liquid are relatively free to move with respect to one another, while the particles in a solid are restricted to small vibrations about points in the crystal lattice. This makes the particle structure of solids much easier to study experimentally than that of liquids. The technique of x-ray diffraction can be applied to find the basic structural unit of the crystal, the unit cell; x-ray diffraction patterns for liquids are diffuse and difficult to interpret.

The particles in a solid tend to pack as closely as possible. (Why?) Most of the metallic elements crystallize in one of the two closest-packed arrangements (cubic or hexagonal). Another common type of packing is body-centered cubic, where the fraction of empty space is only a little greater. In ionic crystals, where the two kinds of ions differ in size, there is a greater variety of packing patterns. Frequently, the larger anions form a close-packed array, slightly expanded if necessary to accommodate the smaller cations fitting into "holes" in the anionic framework.

Much of this chapter is spent in discussing transitions from one state of matter to another. The nature of the equilibria involved can perhaps best be summarized by a phase diagram. At any given temperature, a liquid or solid has a characteristic vapor pressure (curves AB and AC, Figure 11.13). For all liquids and solids, vapor pressure rises exponentially with temperature; the general equation is

$$\log_{10} P = \frac{-\Delta H}{(2.30)(8.31)T} + \text{constant}$$

where P is the vapor pressure, T the absolute temperature, and ΔH the enthalpy change for the transition (heat of vaporization or sublimation), in joules.

Two features of liquid-vapor equilibria which are not apparent from Figure 11.13 are the phenomena of boiling and critical behavior. A liquid boils (i.e., vapor bubbles form within the body of the liquid) when its vapor pressure becomes equal to the applied pressure; the normal boiling point is the temperature at which the vapor pressure is one atmosphere. A liquid can be heated in a closed container to temperatures far above its boiling point. Eventually, however, the kinetic energy of the molecules becomes too great for them to remain in the liquid; at this so-called "critical" temperature, the liquid suddenly and spontaneously vaporizes.

Again referring to the phase diagram on p. 273, we see that there is one particular temperature and pressure (the triple point) at which all three states of a substance may exist in equilibrium with each other. For most substances, the triple point pressure is relatively low. For a few substances, including CO_2, the triple point pressure exceeds one atmosphere. Such

148 • 11–Liquids and Solids; Phase Changes

substances, when heated in an open container, pass directly from the solid to the vapor state (sublime).

The line labeled AD in Figure 11.13 describes the equilibrium between the solid and liquid states. To the extent that this line deviates from the vertical, the melting point of the substance changes with pressure. For most substances, where the solid phase is the more dense, an increase in pressure favors the formation of solid and the melting point rises; the line AD inclines to the right. For water, where the liquid is the more dense phase, the reverse is true. The effect is always small; pressures in the range of 50 to 100 atm are required to change the melting point by as much as 1°C.

We need to keep in mind the ideas discussed back in Chapter 5 about the dynamic nature of matter at the atomic-molecular level. For example, the distribution of energies is central to our explanation of vapor pressure and its dependence on temperature. At any given temperature, a fraction of the molecules will have considerably greater than average energy — enough to escape from a liquid surface — while others are left waiting for collisions to impart the extra energy required to overcome intermolecular forces. Again, as the temperature is increased, a greater fraction of the molecules possess energy sufficient to break away from the rest of the condensed state, so that vapor pressure increases exponentially. You can now see, at least qualitatively, how for a given substance (i.e., for a given magnitude of intermolecular forces, as measured by ΔH_{vap} or by the surface tension) the temperature determines the vapor pressure.

BASIC SKILLS

1. **Have sufficient understanding of the concept of vapor pressure to work problems such as 11.1, 11.10, 11.11, 11.29, and 11.30.**

This is an awkward way to phrase a "skill," but the only alternative is to resort to a lot of jargon. In certain parts of these problems, you have to apply the gas laws to calculate pressures or volumes (see also Example 11.1). You should keep in mind, however, that these laws can be applied only when there is a single, gaseous phase present. They cannot, for example, be used to calculate the pressure of a vapor in equilibrium with its liquid. That pressure is a characteristic physical property of a particular liquid at a given temperature, known as its equilibrium vapor pressure.

Pursuing this idea a bit further, a vapor at a pressure above its equilibrium vapor pressure is unstable; it will spontaneously condense until the pressure drops to the equilibrium vapor pressure. We cannot, for example, maintain steam at 100°C at a pressure above 1 atm. On the other hand, steam at 100°C at a pressure below 1 atm will remain at that pressure

Basic Skills • 149

indefinitely, provided there is no liquid water around to establish equilibrium.

2. Use the Clausius-Clapeyron equation (Equation 11.4) to calculate one of the quantities P_2, T_2, or ΔH_{vap} given the values of the other two and P_1 at T_1.

This skill is illustrated in Example 11.2. Here, the desired quantity is P_2, given T_2 (60°C), ΔH_{vap} (30 500 J/mol), P_1 (36.3 kPa), and T_1 (50°C). Problem 11.13a is analogous to Example 11.2; in 11;13b, you are required to calculate the temperature at which the vapor pressure is 101.3 kPa. Finally, in Problems 11.2 and 11.32, you are given two temperatures and a ratio of vapor pressures and asked to calculate a heat of vaporization.

Notice that in applying Equation 11.4, you must

– have at least a nodding aquaintance with logarithms (and anti-logarithms yet!);
– express temperatures in K;
– use or obtain ΔH in J/mol;
– express both pressures in the same units.

3. **Given the vapor pressure of a liquid at two or more temperatures, use Equation 11.3 to determine the heat of vaporization by a graphical method.**

This problem perhaps arises most frequently as a laboratory exercise. To illustrate the principle involved, consider how we might determine the heat of vaporization of water from Figure 11.5, p. 261. To obtain the slope of the straight line, we may choose any two points. A convenient choice might be to pick the intersections of the line with the horizontal lines labelled 2.0 and 0.0 respectively. The x values corresponding to these intersections appear to be about "2.70" and "3.60" respectively. In other words:

$$y_2 = 2.00; \; 10^3 x_2 = 2.70; \; x_2 = 2.70 \times 10^{-3}$$

$$y_1 = 0.00; \; 10^3 x_1 = 3.60; \; x_1 = 3.60 \times 10^{-3}$$

$$\text{Slope} = \frac{y_2 - y_1}{x_2 - x_1} = \frac{2.00 - 0.00}{(2.70 - 3.60)10^{-3}} = \frac{2.00}{-0.90} \times 10^3 = -2.2 \times 10^3$$

but, slope = $\dfrac{\Delta H_{vap}}{(2.30)8.31}$; $\Delta H_{vap} = -(2.30)(8.31)$ slope
$= -(2.30)(8.31)(-2.2 \times 10^3)$
$= 42\,000$ J

150 • 11–Liquids and Solids; Phase Changes

Actually, this example is simplified by the fact that the plot has already been made for us. In Problems 11.12 and 11.31, you have to make your own plot and then carry out the type of analysis we have just gone through.

4. Use Trouton's Rule (Equation 11.5) to relate the molar heat of vaporization of a liquid to its normal boiling point.

Given that diethyl ether boils at 35°C at 1 atm pressure, use Trouton's Rule to estimate its molar heat of vaporization. _____

From Equation 11.5, we see that $\Delta H_{vap} \approx 88 \times T_b$. Noting that T_b must be in K, we have:

$$T_b = (35 + 273)K = 308K$$

$$\Delta H_{vap} = (88 \times 308) \text{ J} = 27\ 100 \text{ J/mol}$$

The measured value (Table 11.1) is 25 900 J/mol.

Problem 11.14 is entirely analogous to the example just worked. In Problem 11.33, one extra step is involved.

5. Use the Bragg equation (Equation 11.6) to relate the angle of diffraction, θ, and the wavelength of x-rays used, λ, to the distance between adjacent planes of atoms, d.

Problems of this type, of which 11.16 and 11.35 are typical, require little more than simple substitution into an equation. However, you will need trig tables or a calculator with trig functions. This is perhaps the only time you will need such functions in the general chemistry course.

6. **Given the type of unit cell (simple cubic, face-centered cubic, or body-centered cubic), relate the cell dimensions to such quantities as atomic radius and density.**

The type of calculation involved here is illustrated, for a face-centered cubic unit cell, in Example 11.3. Problems 11.3 and 11.37 are entirely analogous to that example; Problem 11.18 applies the same reasoning to a body-centered cubic unit cell. The various relations involved in three types of unit cells are summarized in the following table.

Basic Skills • 151

	Simple	Face-Centered	Body-Centered
no. of atoms per cell	1	4	2
atoms touch along:	edge	face diagonal	body diagonal
relation between atomic radius (r) and cell length (ℓ)	$2r = \ell$	$4r = \ell\sqrt{2}$	$4r = \ell\sqrt{3}$

All the relationships just discussed apply to crystals in which only one kind of particle (e.g., a metal atom) is present. The same kind of geometric reasoning can be applied to ionic crystals, where we are dealing with two different kinds of particles. This application is covered in Problems 11.19 and 11.38.

7. **Given the phase diagram for a pure substance, decide what phases are present at a given temperature and pressure.**

This skill is required to work Problems 11.4, 11.22, and 11.41. To understand what it involves, consider the phase diagram for water (Fig. 11.13, p. 273). Note that within any given area such as that labelled "vapor" in Figure 11.13, only one phase is present. Along any line in the diagram, such as AB, two phases are in equilibrium. At the triple point, A, all three phases are present.

Referring to Figure 11.13, what phases are present at B?_____ at 25°C and 1 kPa?_____ at C?_____ at −5°C and 2 kPa?_____

Since B lies on the vapor pressure curve of the liquid, the two phases, liquid and vapor, must be in equilibrium. Similarly, the two phases solid and vapor must be in equilibrium at C. Locating 25°C and 1 kPa on the diagram, you should find that it is entirely within the area labelled "vapor"; only vapor is present. The point at −5°C and 2 kPa clearly lies within the solid region; only ice is present.

---

8. **Using data such as that given in Table 11.3, make calculations regarding ΔH for phase changes.**

This skill is illustrated in Example 11.4 and in the following example. Note that in all problems of this type it is assumed that the heat absorbed is numerically equal to that evolved — i.e., no heat is lost to the surroundings.

---

152 • 11–Liquids and Solids; Phase Changes

20.0 g of steam at 100°C is condensed in one kilogram of water, originally at 25°C. What is the final temperature? _____

Heat is evolved when the steam condenses (2257 J/g) and when it cools from 100°C to the final temperature, t [4.18 J/(g · °C)].

$$Q_{steam} = -2257 \frac{J}{g} \times 20.0 \text{ g} + 4.18 \frac{J}{g°C} \times 20.0 \text{ g} \times (t - 100°C)$$

$$= -45\,100 \text{ J} + 83.6 \frac{J}{°C} (t - 100°C)$$

Heat is absorbed by the water as it is warmed from 25°C to t:

$$Q_{water} = 4.18 \frac{J}{g°C} \times 1000 \text{ g} \times (t - 25°C) = 4180 \frac{J}{°C} (t - 25°C)$$

Since there is no net loss of heat: $Q_{steam} + Q_{water} = 0$ or: $Q_{water} = -Q_{steam}$

Solving: $4180(t - 25°C) = 45\,100 - 83.6(t - 100°C)$

$4260t = 158\,000;\ t = 37°C$

See Problems 11.5, 11.23, 11.24, 11.42, and 11.43.

SELF-TEST

True or False

1. At 50°C, the vapor pressure of liquid A is found to be ( )
5 kPa; that of liquid B, 40 kPa. One can be reasonably sure that liquid A has the higher normal boiling point.

2. For these same two liquids, one can state that the liquid ( )
with the higher surface tension at 50°C is probably A.

3. For any pure substance, the melting point is always a little ( )
above the freezing point.

4. Not all solids are crystalline. ( )

5. Whenever bubbles form within a sample of liquid, the liquid ( )
is said to be boiling.

6. Generally, when a solid melts to form a liquid, the density ( )
decreases.

7. For all solids and liquids, vapor pressure increases linearly ( )
with temperature.

8. Below the triple point temperature, only solid can exist. ( )

9. Pure water cannot exist as a liquid below 0°C. ( )

10. The heat of vaporization of a given substance is generally ( )
larger than the heat of fusion.

Multiple Choice

11. When one mole of liquid water is placed in a 100 cm³ flask ( )
at 25°C, it eventually establishes a constant pressure of 3.2 kPa. If
a 200 cm³ flask were used instead, the final pressure would be
    (a) 1.6 kPa   (b) 3.2 kPa
    (c) 6.4 kPa   (d) 1 atm

12. Which substance is likely to have the highest vapor pressure ( )
at any given temperature?
    (a) C_2H_6 (b) C_3H_8
    (c) C_4H_{10} (d) C_5H_{12}

13. The vapor pressure of CCl_4 at 77°C is 1.00 atm. The heat ( )
of vaporization is approximately
    (a) 6.8 kJ   (b) 15.9 kJ
    (c) 30.8 kJ   (d) 39.1 kJ

14. The boiling point of any liquid is: ( )
    (a) 100°C
    (b) the temperature at which as many molecules leave the liquid as return to it
    (c) the temperature at which the vapor pressure is equal to the external pressure
    (d) the temperature at which no molecules can return to the bulk of the liquid

15. The triple point is ( )
    (a) the temperature above which the liquid phase cannot exist
    (b) usually found at a temperature very close to the normal boiling point
    (c) the temperature at which the vapor pressure of a liquid is three times the value at 25°C
    (d) the value of temperature and pressure at which solid, liquid, and gas may exist in equilibrium

16. A certain substance, X, has a triple point temperature of ( )
20°C at a pressure of two atmospheres. Which one of the following

154 • 11–Liquids and Solids; Phase Changes

statements *cannot* be true?
- (a) X can exist as a liquid above 20°C
- (b) X can exist as a solid above 20°C
- (c) liquid X is stable at 25°C and at 1 atm
- (d) both liquid X and solid X have the same vapor pressure at 20°C

17. A certain solid sublimes at 25°C and one atmosphere ( ) pressure. This means
    - (a) the solid is more dense than the liquid
    - (b) the pressure at the triple point is greater than one atmosphere
    - (c) the solid is less dense than the liquid
    - (d) the pressure at the triple point is less than one atmosphere

18. The melting point of benzene at one atmosphere is 5.5°C. ( ) The density of liquid benzene is 0.90 g/cm^3; that of the solid is 1.0 g/cm^3. At an applied pressure of 10 atm, the melting point of benzene
    - (a) is equal to 5.5°C
    - (b) is slightly greater than 5.5°C
    - (c) is slightly less than 5.5°C
    - (d) cannot be estimated from the information given

19. Dry ice, solid CO_2, is frequently used as a refrigerant; it ( ) undergoes sublimation at or near atmospheric pressure. This direct conversion of solid to gas involves an absorption of energy, energy that is required to overcome
    - (a) covalent bonds
    - (b) polar covalent bonds
    - (c) dipole forces
    - (d) dispersion forces

20. When a solid melts ( )
    - (a) greater attractive forces appear
    - (b) the molecules become more randomly oriented
    - (c) the molecules become less randomly oriented
    - (d) the container warms up

21. The number of closest neighbors (making contact) in a ( ) body-centered cubic lattice of identical spheres is
    - (a) 4
    - (b) 6
    - (c) 8
    - (d) 12

22. A certain metal crystallizes in a face-centered cubic unit ( ) cell. The relationship between the atomic radius (r) of the metal and the length (ℓ) of one edge of the cube is:
    - (a) $2r = \ell$
    - (b) $2r = \sqrt{2}\ell$
    - (c) $4r = \sqrt{2}\ell$
    - (d) $4r = \sqrt{3}\ell$

23. A chemistry handbook usually gives the pressure at which ( ) the bp of a pure substance is measured but generally does *not* indicate

the pressure for the mp. Why?
- (a) both melting point and boiling point are always measured at the same pressure
- (b) the melting point is usually nearly independent of pressure
- (c) solids are more often impure and therefore have variable melting points
- (d) all melting points are measured at one atmosphere

24. A sample of gaseous H_2O at a pressure of 2.34 kPa is in equilibrium with liquid H_2O at 20°C. On slowly reducing the volume available to the system, you observe that the pressure remains at 2.34 kPa. The observation is accounted for by assuming that   ( )
- (a) gaseous water doesn't behave ideally under these conditions
- (b) some vapor escapes from the container
- (c) your pressure measurement was incorrect; it should have been higher
- (d) some vapor condenses to form liquid

25. For a liquid in equilibrium with its vapor, a straight line is obtained by plotting   ( )
- (a) P vs. T
- (b) log P vs. T
- (c) log P vs. 1/T
- (d) 1/P vs. log T

26. Reducing the pressure on a liquid with a vacuum pump can cause   ( )
- (a) boiling
- (b) freezing
- (c) evaporation
- (d) all of these

27. A certain metal fluoride crystallizes in such a way that fluorine atoms occupy simple cubic lattice sites, while metal atoms occupy the body centers of half the cubes. The formula of the metal fluoride is   ( )
- (a) M_2F
- (b) MF
- (c) MF_2
- (d) MF_8

28. Which phase change may be taken advantage of in the purification of a chemical substance?   ( )
- (a) freezing
- (b) boiling
- (c) sublimation
- (d) any of these

29. The heat of vaporization of NaCl is much larger than the heat of vaporization of H_2O. This is a result of NaCl having   ( )
- (a) a higher formula mass
- (b) stronger forces of attraction between units of structure
- (c) a higher density
- (d) a higher boiling point

156 • 11–Liquids and Solids; Phase Changes

30. A certain metal crystallizes in a simple cubic structure. At a ( )
certain temperature, it rearranges to give a body-centered structure.
In this transition, the density of the metal will
   (a) decrease
   (b) increase
   (c) remain unchanged
   (d) change in an unknown way

31. The critical temperature of N_2 is 124 K. If one wishes to ( )
liquefy a sample of N_2, one needs to
   (a) increase pressure at constant temperature
   (b) decrease pressure at constant temperature
   (c) increase temperature at constant pressure
   (d) decrease temperature at constant pressure

32. The triple point of iodine is 12 kPa and 115°C. This means ( )
that $I_2(l)$
   (a) is more dense than $I_2(s)$
   (b) cannot exist above 115°C
   (c) cannot exist at 1 atm pressure
   (d) cannot have a vapor pressure less than 12 kPa

33. Consider the phase diagram ( )
sketched here. What would be observed
when the temperature and pressure are
changed from Point 3 to Point 1?
   (a) freezing
   (b) melting
   (c) boiling
   (d) combustion

34. The number of atoms assigned to a unit cell of a face- ( )
centered cubic lattice of identical atoms is
   (a) 1   (b) 2
   (c) 4   (d) 14

35. Most crystalline solids can be expected to ( )
   (a) be simple cubic   (b) have numerous defects
   (c) be metals   (d) have semiconductivity

Problems

36. A certain liquid has a normal boiling point of 50°C.
   (a) Estimate the heat of vaporization, using Trouton's Rule.
   (b) Using the heat of vaporization calculated in (a), estimate the
       vapor pressure of the liquid at 25°C.

37. One mole of $H_2O(l)$ is introduced into an evacuated 100 cm^3 container at $27°C$. At this temperature, it establishes its equilibrium vapor pressure of 3.57 kPa.
   (a) What will be the pressure inside the container if the volume is increased to 200 cm^3 at $27°C$?
   (b) What will be the pressure inside the container if the temperature is raised to $127°C$? $\Delta H_{vap} = 41.8$ kJ/mol.

38. Silver (AM = 108) crystallizes in a face-centered cubic structure with the edge of the unit cell being 0.407 nm.
   (a) What is the atomic radius of silver?
   (b) How many atoms are there per unit cell?
   (c) What is the density of silver metal?

*39. Lithium iodide crystallizes with the structure of NaCl. The shortest Li-I internuclear distance has been determined by x-ray diffraction to be 0.302 nm.
   (a) Calculate the shortest I-I distance.
   (b) What assumptions must be made for this calculated distance between iodine nuclei to be equated to the diameter of an iodine atom (I^-) in LiI?
   (c) Assuming indeed that the distance calculated in (a) is the diameter of the iodine atom in LiI, what can be said about the size of the lithium atom (Li^+)? The tabulated radius for Li^+ is 0.060 to 0.068 nm, depending on where you look. Is it consistent with your conclusions?

[This problem illustrates one of the few methods for determining ionic radii.]

*40. Iron(II) sulfide, which occurs as the mineral "magnetic pyrites," is normally deficient in iron. That is, instead of equal numbers of Fe^{2+} and S^{2-} ions, there are fewer Fe^{2+} atoms and some of the iron atoms present are actually Fe^{3+}. (Overall, there is still electrical neutrality.) For a sample of pyrites of composition $Fe_{0.86}S$, how many such defects are there in the iron sulfide structure? In particular, what is the ratio of Fe^{3+} to Fe^{2+} in the sample?

SELF-TEST ANSWERS

1. **T** (At $50°C$, B is closer to the boiling point than A since its vapor pressure is closer to 1.00 atm.)
2. **T** (Both a lower vapor pressure and a higher surface tension would result from stronger intermolecular attractions.)
3. **F** (They are one and the same equilibrium temperature.)
4. **T** (Many solids, such as glass and rubber, are *amorphous*, lacking

158 • 11–Liquids and Solids; Phase Changes

the regularity of atomic placement found in crystals.)
5. **F** (Bubbles often indicate dissolved gases coming out of solution – Chapter 12.)
6. **T** (Water is one of the few exceptions.)
7. **F** (Exponentially.)
8. **F** (What about the pressure?)
9. **F** (Supercooling may occur. Also consider higher pressures.)
10. **T** (Why? Answer in terms of intermolecular forces – Chapter 9.)
11. **b** (Could all the liquid have evaporated? Check by using the Ideal Gas Law, Chapter 5. Vapor pressure depends only on temperature.)
12. **a** (Smallest of these nonpolar molecules, weakest dispersion forces – Chapter 9.)
13. **c** (Note that 1.00 atm means this is the bp; apply Trouton's Rule.)
14. **c** (If you chose b, consider the boiling of water in an open container.)
15. **d** (And is usually very close to the melting point. Why?)
16. **c** (Construct a phase diagram.)
17. **b**
18. **b** (The solid-liquid line must incline to the right.)
19. **d** (Chapter 9.)
20. **b**
21. **c** (Contact with the body-centered sphere is made along the body diagonals.)
22. **c** (Contact is made along a face diagonal.)
23. **b** (The solid-liquid line is usually almost vertical.)
24. **d** (Equilibrium is maintained.)
25. **c**
26. **d** (Evacuation will not only lower the pressure, at least temporarily, but also remove the "hottest" molecules and thus lower the temperature.)
27. **c** (One F atom per cube; $\frac{1}{2}$ M atom per cube.)
28. **d** (See the discussion in Chapters 1 and 11.)
29. **b** (What forces are overcome, what structural units are separated? Chapter 9.)
30. **b** (The second arrangement is the more compact.)
31. **d** (It cannot be liquefied at temperatures above 124 K, regardless of the pressure.)
32. **d** (The liquid-vapor curve ends there.)
33. **a** (What phases are present at the other points? What occurs when you go from Point 3 to Point 2?)
34. **c** $\left(\frac{1}{8}\text{ of each corner atom} + \frac{1}{2}\text{ of each face atom.}\right)$
35. **b**

Self-Test Answers • 159

36. (a) $88 \times 323 = 28\,400$ J/mol
    (b) Note that vapor pressure is 101.3 kPa at 50°C:
    $$\log \frac{101.3}{P} = \frac{28\,400(25)}{(2.30)(8.31)(298)(323)} = 0.386; \quad \frac{101.3}{P} = 2.43;$$
    $P = 41.7$ kPa

37. (a) 3.57 kPa (independent of volume)
    (b) $\log \dfrac{P}{3.57} = \dfrac{41\,800(100)}{(2.30)(8.31)(300)(400)} = 1.82;\quad \begin{aligned}P &= 66(3.57)\\ &= 240 \text{ kPa}\end{aligned}$

38. (a) $\dfrac{0.407 \text{ nm} \times \sqrt{2}}{4} = 0.144$ nm
    (b) 4
    (c) $V_{cell} = (4.07 \times 10^{-8} \text{ cm})^3 = 6.74 \times 10^{-23}$ cm³
    mass of four atoms $= 4 \times 108$ g$/6.02 \times 10^{23} = 71.8 \times 10^{-23}$ g
    density $= 71.8$ g$/6.74$ cm³ $= 10.7$ g/cm³

*39. (a) distance (I-I) $= 0.302 \times \sqrt{2}$ nm $= 0.427$ nm
    (b) If 0.427 nm is equated to $2 \times I^-$ radius, then one must assume I-I contact.
    (c) If $r_{I^-} = 0.214$ nm, then r_{Li^+} cannot exceed the value calculated from
    $$r_{I^-} + r_{Li^+} \leq 0.302 \text{ nm}$$

    That is, r_{Li^+} is less than or equal to 0.088 nm.
    Apparently there is anion-anion contact.

*40. Consider one mole of $Fe_{0.86}S$:
    moles Fe atoms = moles Fe^{2+} + moles Fe^{3+} = 0.86
    total positive charge = total negative charge, or

    $3 \times$ moles Fe^{3+} + $2 \times$ moles Fe^{2+} = 2 mol of positive charge

    This gives us two equations in two unknowns:
    $$0.86 = n_{Fe^{3+}} + n_{Fe^{2+}}$$
    $$2 = 3n_{Fe^{3+}} + 2n_{Fe^{2+}}$$

    Simultaneous solution gives moles $Fe^{2+} = 0.58$, moles $Fe^{3+} = 0.28$, or atom ratio $Fe^{3+}/Fe^{2+} = 0.48$
    (Some 33% of the iron atoms are Fe^{3+}.)

SELECTED READINGS

Liquids, their structure and properties, are discussed in:

Apfel, R. E., The Tensile Strength of Liquids, *Scientific American* (December 1972), pp. 58–71.
Bernal, J. D., The Structure of Liquids, *Scientific American* (August 1960), pp. 124–128.
Reid, R. C., Superheated Liquids, *American Scientist* (March-April 1976), pp. 146–156.
Turnbull, D., The Undercooling of Liquids, *Scientific American* (January 1965), pp. 38–46.

Solids, crystalline and glass, are discussed in the books by Holden and Moore listed for Chapter 9 and:

Darragh, P. J., Opals, *Scientific American* (April 1976), pp. 84–95.
Fullman, R. L., The Growth of Crystals, *Scientific American* (March 1955), pp. 74–80.
Greene, C. H., Glass, *Scientific American* (January 1961), pp. 92–105.
McQueen, H. J., The Deformation of Metals at High Temperatures, *Scientific American* (April 1975), pp. 116–125.
Sanderson, R. T., The Nature of "Ionic Solids," *Journal of Chemical Education* (September 1967), pp. 516–523.

Visualizing the third dimension and experimentally probing it:

Bragg, L., X-Ray Crystallography, *Scientific American* (July 1968), pp. 58–70.
Kapechi, J. A., An Introduction to X-Ray Structure Determination, *Journal of Chemical Education* (April 1972), pp. 231–236.
Wells, A. F., *The Third Dimension in Chemistry*, New York, Oxford University Press, 1956.

12
SOLUTIONS

QUESTIONS TO GUIDE YOUR STUDY

1. What kinds of processes occur within solids? Within liquids? For example, what happens at the molecular level when two liquids are mixed, or when a solid dissolves in a liquid?

2. How do the properties of a substance (such as mp, bp) change when a second substance is dissolved in it?

3. Can you think of some naturally occurring solutions? (Recall that we generally do not encounter pure substances outside the lab.)

4. How do you quantitatively describe the composition of a solution?

5. Are there simple laws relating the properties of a solution to the properties of solute and solvent? How do the properties of a solution depend on the relative amounts of its components?

6. What properties of substances determine the extent to which one will dissolve in the other? What generalizations can be made?

7. How do changes in conditions, such as temperature and pressure, affect solubility?

8. How would you show experimentally that a given solution is saturated? Supersaturated? How would you prepare such solutions?

9. Can you describe the particle structure, and the intermolecular forces, of solutions? Can you account for any energy effects accompanying the formation of a solution?

10. To what uses are solutions put in the laboratory?

11.

12.

YOU WILL NEED TO KNOW

Concepts

1. How to predict the geometry and polarity of molecules, bond type, and the kinds and relative magnitudes of intermolecular forces – Chapters 8, 9.
2. The meaning of partial pressure – Chapter 5.

Math

1. How to express amounts of substances in the various common units: grams, moles, number of particles – Chapter 3.

CHAPTER SUMMARY

We saw in Chapter 9 that the magnitude of a substance's intermolecular forces determines whether it exists as a gas, liquid, or solid at normal temperatures and pressures (e.g., 25°C, 1 atm). Again, we might expect two substances with intermolecular forces of about the same magnitude to be infinitely soluble in each other. Putting these two generalizations together, we expect complete miscibility only if the two substances involved are in the same physical state. Experiment confirms this reasoning: all gases are infinitely soluble in one another; complete miscibility is the rule rather than the exception among liquids of similar polarity. In contrast, solids and gases always show limited solubility in liquid solvents. Carrying this reasoning one step further, we deduce that the closer a solid or gas is to the liquid state (i.e., the lower the melting point of the solid or the higher the normal boiling point of the gaseous substance) the greater will be its solubility. At the opposite extreme, solutes whose intermolecular forces differ vastly from those of the solvent will be virtually insoluble in it (e.g., the permanent gases and hydrocarbons in water).

It is, of course, possible to change the solubility of a solute by changing the external conditions of temperature and pressure. (*Solubility* refers to the extent of a particular kind of reaction *at equilibrium*. In Chapter 11, we discussed the effect of changes in temperature and pressure on a system at equilibrium.) If the solution process is endothermic, an increase in temperature promotes solubility. This is almost always the case with solid-liquid systems and usually true when both components are liquids. With a gas, the solution process may be either endothermic or exothermic;

the solubility of the permanent gases in organic solvents normally increases with temperature while the same gases become less soluble in water when the temperature is raised. Solubility is little affected by pressure except for gas-liquid systems where it is directly proportional to the partial pressure of the gas over the solution.

Various methods are used to express the concentrations of solutions — i.e., the relative amounts of solute and solvent. Frequently, the mass fractions are given by quoting mass per cents of the components or, for very dilute solutions, "parts per million" or even "parts per billion" of solute (1 ppm = 1 g solute/10^6 g solvent). In the laboratory, reagent concentrations are most frequently expressed in terms of number of moles per cubic decimetre (M). Two other concentration units, mole fraction and molality, are particularly useful for expressing the colligative properties of solutions. The terminology here is unfortunate; students often confuse molality (m) with "molarity" (M), or even with morality, which is something else altogether.

Almost without exception, the addition of a small amount of solute lowers the freezing point of a solvent; if the solute is nonvolatile, the boiling point will be raised as well. Both of these effects can be related to the lowering of solvent vapor pressure that always accompanies the formation of a solution. One can also explain the phenomenon of osmosis in terms of vapor pressure lowering: water or other solvent moves through a semipermeable membrane from a region of high vapor pressure (pure solvent) to one of low vapor pressure (solution). Osmosis can be prevented by exerting sufficient force on the solution side of the membrane; the pressure required to do this is referred to as the osmotic pressure.

Each of the effects just described (vapor pressure lowering, freezing point lowering, boiling point elevation, osmotic pressure) is a colligative property; its magnitude depends primarily upon the concentration of solute particles rather than their type. The appropriate equations appear in the text (Equations 12.6–12.10). Notice that these equations incorporate a constant of proportionality which in every case but one is characteristic of the particular solvent. The exception is the expression for osmotic pressure, where the gas law constant R appears. (This constant keeps popping up in the most unexpected places! How would you rationalize the similarity of the two equations $\pi V = nRT$ and the Ideal Gas Law? In what ways is a dilute solution like a gas?) These equations can be used in a very practical way to calculate vapor pressures, freezing points, boiling points, and osmotic pressures of solutions. In addition they represent methods for the experimental determination of the molecular mass of a solute.

BASIC SKILLS

1. Relate the masses of the components of a solution to:
   a. the mass percentages of the components:
   b. the mole fractions of the components;
   c. the molality of the solute.

The application of these skills is illustrated in Examples 12.1 (mass percentages), 12.2 (mole fractions), and 12.3 (molality). Note that each example is worked directly from the defining equation for the concentration unit.

To practice these skills, try Problems 12.1, 12.9a, 12.10b, c, d, 12.28a, b, and 12.29a, b, c. Other parts of these problems involve volume concentrations (M) or, in one case, the use of the Ideal Gas Law.

2. **Given or having calculated two of the three quantities—M, moles of solute, and volume of solution—obtain the other quantity.**

This skill is shown in Example 12.4. In part (a), you are given the volume of solution and the value of M and must calculate first the number of moles of solute and then the number of grams. In part (b), the quantity required is the volume of solution; the number of moles and M are given.

Simple calculations involving M as a concentration unit are required in Problems 12.2a, 12.10a, 12.11, and 12.30. These problems are analogous to Example 12.4 in that you need only substitute into the defining equation to obtain the desired quantity. Problem 12.29d is slightly more involved in that you must first obtain the volume of solution (note that the density is given).

3. **Relate the volumes and concentrations of solutions prepared by dilution with solvent.**

No new skill is involved here. All you need do is recall the definition of M and realize that diluting a solution with solvent does not change the number of moles of solute. Refer to Example 12.5 and Problems 12.2b, 12.12, and 12.28c. In Problem 12.31, you are asked to derive a general equation which could be used to solve any problem of this type.

4. **Use Henry's Law (Equation 12.5) to relate the solubility of a gas in a liquid to its partial pressure.**

This skill is illustrated in Example 12.6 and is required to solve Problems 12.3, 12.14, and 12.33.

5. **Predict the relative solubilities of two different solutes in a given solvent, or of a given solute in two different solvents.**

The principle involved here, that solubility is enchanced by similarity in the type and strength of intermolecular forces, is discussed in the text and illustrated by the following example.

At 25°C and 1 atm, which would be the more soluble in a nonpolar liquid solvent (such as benzene), $H_2O(l)$ or $C_6H_{14}(l)$?_____ $H_2(g, bp = -253°C)$ or $CH_4(g, bp = -183°C)$? _____ Naphthalene (s, mp = 80°C) or anthracene (s, mp = 218°C)? _____

In the first case, the obvious choice is hexane, C_6H_{14}, since its intermolecular forces are of the same type (dispersion) as those in benzene. Water, which is hydrogen-bonded, is only very slightly soluble in nonpolar solvents. With the other pairs, the type of intermolecular force (dispersion) is the same in both cases. The choice for the more soluble species is the solute whose intermolecular forces are nearest in *magnitude* to those of the liquid solvent. These are the gas, CH_4, and the solid, naphthalene, which are closest to the liquid state at room temperature. Gases which are difficult to condense (H_2) and solids which are difficult to melt (anthracene) have intermolecular forces quite different in magnitude from those of a liquid such as benzene, and hence have relatively low solubilities.

Problems 12.15, 12.16, 12.34, and 12.35 require the use of this skill; in certain parts of these problems you are expected to recognize the physical state of the solute.

6. **Use Raoult's Law to relate:**
   a. the vapor pressure of solvent in a solution to its mole fraction;
   b. the vapor pressure lowering to the mole fraction of solute;
   c. the total vapor pressure of a solution to the mole fractions of solute and solvent.

The appropriate equations for these calculations are given in the text. The calculations for (b) and (c) are illustrated in Example 12.7 and the paragraph that follows. To illustrate (a), consider the following example.

What is the vapor pressure of water in a solution containing 100 g of sugar, $C_{12}H_{22}O_{11}$, in 500 g of water, H_2O, at 25°C? _____

In order to apply Equation 12.6 we need to know the vapor pressure of pure water at 25°C (3.17 kPa) and the mole fraction of water in the solution.

Since the molecular masses of H_2O and $C_{12}H_{22}O_{11}$ are 18.0 and 342, respectively:

$$X_1 = \frac{\text{no. moles } H_2O}{\text{no. moles } H_2O + \text{no. moles } C_{12}H_{22}O_{11}} = \frac{500/18.0}{(500/18.0) + (100/342)} = 0.990$$

$$P_1 = X_1 P_1{}^0 = (0.990)(3.17 \text{ kPa}) = 3.14 \text{ kPa}$$

Problems 12.17, 12.18, 12.36, and 12.37 all apply Raoult's Law to vapor pressure calculations.

7. Use Equations 12.8 (boiling point elevation) and 12.9 (freezing point lowering) to obtain:
    a. the boiling point or freezing point of a nonelectrolyte solution;
    b. the molecular mass of a nonelectrolyte.

The calculations involved in (a) are illustrated in Example 12.8; those required to obtain the molecular mass of a solute are shown in Example 12.9. Note that:
— in order to work problems of this type, you must be thoroughly familiar with the concept of molality, the concentration unit used in Equations 12.8 and 12.9.
— these equations apply only to nonelectrolytes (i.e., molecular solutes); electrolyte solutions are considered in Chapter 13.
Problems 12.19 and 12.38 require skill 7(a); Problems 12.24 and 12.43 apply skill 7(b). Problems 12.20 and 12.39 are similar but involve additional conversions.

8. Use Equation 12.10 to relate the osmotic pressure of a solution, π, to concentration, M, or to calculate the molecular mass of the solute.

See Example 12.10 and Problems 12.21, 12.25, 12.40, and 12.44. Note that in Problem 12.25, the solution is sufficiently dilute so that you can assume M to be equal to m.

SELF-TEST

True or False

1. In a solution containing equal numbers of grams of benzene ( )
(MM = 78) and toluene (MM = 92), the mole fractions are each 0.50.

2. A supersaturated solution of air in water could be prepared ( ) by bubbling air through water at 60°C and cooling to 25°C.

3. The mole fractions of all the components of a solution add ( ) to unity.

4. If water saturated with nitrogen at 1 atm is exposed to air, ( ) N_2 will come out of solution.

5. DDT, $C_{14}H_9Cl_5$, should be more soluble in alcohol than in ( ) carbon tetrachloride.

6. The solubility of a solid in a liquid is ordinarily directly ( ) proportional to the absolute temperature.

7. Raoult's Law always applies to the solute in a solution as ( ) well as to the solvent.

8. The boiling point of a one molal water solution of a ( ) nonelectrolyte will be 100.52°C, provided the solution is ideal.

9. Freezing point lowering, boiling point elevation, and ( ) osmotic pressure can all be correlated with vapor pressure lowering.

10. The freezing point of a 1 m solution of K_2SO_4 would be ( ) about the same as that of a 1 m solution of urea.

Multiple Choice

11. When one mole of KCl is dissolved in a kilogram of water, ( ) the concentration of Cl^- is
 (a) 0.5 m (b) 0.5 M
 (c) 1.0 m (d) 1.0 M

12. A student wishes to prepare 100 cm³ of 0.50 M NaCl from ( ) 2.00 M NaCl. What volume of the more concentrated solution should he start with?
 (a) 25 cm³ (b) 50 cm³
 (c) 100 cm³ (d) 400 cm³

13. If you were asked to prepare a 0.30 M NaCl solution, which ( ) of the following stock solutions of NaCl would you *not* use?
 (a) 0.20 M (b) 0.40 M
 (c) 0.60 M (d) 1.0 M

14. The solubility of a certain salt in water is 22 g/dm³ at 25°C ( ) and 60 g/dm³ at 80°C. A student prepares 500 cm³ of saturated solution at 80°C, cools it to 25°C, and adds a tiny crystal of salt. How many grams of salt come out of solution?

168 • 12–Solutions

    (a) none   (b) 11
    (c) 19   (d) 38

15. The number of grams of Na_2SO_4 (FM = 142) required to ( ) prepare 2.00 dm^3 of 1.50 M solution is
    (a) 3   (b) 213
    (c) 142   (d) 426

16. To obtain 12.0 g of K_2CrO_4 from a solution labeled "5.0% ( ) K_2CrO_4 by mass," how many grams of solution should you weigh out?
    (a) 2.4   (b) 5.0
    (c) 12   (d) 240

17. Which of the following has the least effect on the solubility ( ) of a solid in a liquid solvent?
    (a) composition of solute   (b) composition of solvent
    (c) temperature   (d) pressure

18. The solubility of CO_2 in water at 20°C is 3.4 g/dm^3 at 1 ( ) atm. If the CO_2 pressure is raised to 3 atm, the solubility (g/dm^3) is expected to be
    (a) 3.4/3   (b) 3
    (c) 3.4   (d) 3.4 × 3

19. For which of the following pairs would you expect ( ) solubility to be the greatest?
    (a) O_2-water   (b) sugar-water
    (c) sugar-benzene   (d) O_2-N_2

20. Which one of the following liquids would you expect to be ( ) most soluble in water?
    (a) CH_3COOH (b) C_2H_5-O-C_2H_5
    (c) CCl_4 (d) n-C_7H_{16}

21. A student wants to remove naphthalene, an aromatic ( ) hydrocarbon, from his lab coat. What solvent would you recommend?
    (a) water   (b) ethyl alcohol
    (c) benzene   (d) sulfuric acid

22. When one mole of a nonvolatile nonelectrolyte is dissolved ( ) in two moles of a solvent, the vapor pressure of the solution, relative to that of the pure solvent, is
    (a) 1/3   (b) 1/2
    (c) 2/3   (d) cannot tell

23. At 20°C, ethyl alcohol has a vapor pressure of 6 kPa; ( ) diethyl ether, 60 kPa. What is the expected total pressure over

an equimolar solution of alcohol and ether at 20°C, assuming ideal behavior?
(a) 6 kPa
(b) 33 kPa
(c) 60 kPa
(d) $n_{tot}RT/V$

24. Which one of the following solution properties is not a (   ) colligative property?
(a) freezing point
(b) melting point
(c) osmotic pressure
(d) color

25. The osmotic pressure of a 1 M solution of sugar at room (   ) temperature would be closest to
(a) 25 kPa
(b) 100 kPa
(c) 250 kPa
(d) 2500 kPa

26. To determine the molecular mass of a polymer with an (   ) approximate molecular mass of 10 000, one would probably measure the
(a) osmotic pressure of a solution
(b) boiling point of a solution
(c) density of the solid
(d) vapor pressure of a solution

27. A small amount of a certain solute which melts at 800°C is (   ) added to water. The solution will be expected to freeze
(a) above room temperature
(b) slightly above 0°C
(c) at 0°C
(d) slightly below 0°C

28. In a 0.1 M solution of NaCl in water, which one of the (   ) following will be closest to 0.1?
(a) mole fraction NaCl
(b) mole fraction water
(c) mass % NaCl
(d) molality

29. A solution of 1.00 g of a nonelectrolyte in 20.0 g of water (   ) freezes at −0.50°C. The molecular mass of the nonelectrolyte is
(a) 1.86/(0.50)(0.020)
(b) 1.86/(0.50)(20.0)
(c) 0.50(20.0)/1.86
(d) 0.020(0.50)/1.86

30. Separate solutions are made of each of the following (   ) substances by dissolving 10 g in 1000 g of benzene. Which will start to freeze at the lowest temperature?
(a) CH_2Cl_2
(b) $CHCl_3$
(c) CCl_4
(d) all will freeze at the same T

31. Which of the following properties should be exhibited by a (   ) 0.01 m aqueous solution of a nonvolatile nonelectrolyte?
(a) the solution readily conducts electricity

(b) the temperature drops during freezing
(c) the temperature remains constant during freezing
(d) the boiling point is lowered

Problems

32. A water solution contains 12.0 g of sugar (MM = 342) in 200 g of water; its density is 1.022 g/cm^3. Calculate:
    (a) the mole fraction of sugar
    (b) the molality of sugar
    (c) M sugar

33. When 0.946 g of fructose is dissolved in 150 g of water (k_f = 1.86°C), the resulting solution is found to have a freezing point of −0.065°C.
    (a) What is the molecular mass of fructose?
    (b) If the simplest formula of fructose is CH_2O, what must be its molecular formula?

34. A student dissolved 1.00 g of polystyrene in enough toluene to make 0.100 dm^3 of solution. She measured the osmotic pressure and found it to be 0.261 kPa at 27°C [R = 8.31 kPa · dm^3/(mol · K)]. What was the molecular mass of the polystyrene?

*35. A beaker containing 1.68 g of sugar, $C_{12}H_{22}O_{11}$, in 20.00 g H_2O and another containing 2.45 g of a high-melting nonelectrolyte in 20.00 g H_2O are placed in a bell jar and allowed to come to equilibrium (see Fig. 12.6). The total mass of the sugar solution at equilibrium is 24.90 g. Estimate the MM of the nonelectrolyte.

*36. A student measures the bp of CS_2 (l) to be 46.1°C and that of a 1.00 m solution to be 48.5°C. When 1.5 g of sulfur are dissolved in 12.5 g CS_2, the bp is found to be 47.2°C. Determine the molecular formula of sulfur.

SELF-TEST ANSWERS

1. F (Larger number of moles of benzene.)
2. F (Solubility increases as temperature drops; it wouldn't even be saturated at the lower temperature.)
3. T
4. T (The partial pressure of N_2 above the solution is reduced.)
5. F (The alcohol contains hydrogen bonds, the DDT does not. Dissimilarity in intermolecular forces results in low solubility.)
6. F (The relationship is not linear and depends on ΔH for the solution process.)

Self-Test Answers • 171

7. F (The solute would need to be volatile, and even then it may not obey Raoult's Law.)
8. F (Depends on the applied pressure as well as on solute volatility.)
9. T
10. F (More solute particles with K_2SO_4 — see Chapter 13.)
11. c (Again, Chapter 13 discusses electrolyte solutions in more detail.)
12. a (This will contain the 0.050 mol of NaCl needed.)
13. a (The solution would have to be concentrated — e.g., by evaporation — less easily done, accurately, than diluting one of the others.)
14. c (At 80°C, 500 cm³ of saturated solution contains 30 g salt; at 25°C, 11 g.)
15. d (2.00 dm³ × 1.50 mol/dm³ × 142 g/mol.)
16. d (12.0 g salt × 100 g solution/5.0 g salt.)
17. d (Pressure has a significant effect only when gases are involved.)
18. d (Henry's Law.)
19. d (Same physical state, similar magnitudes and types of intermolecular forces.)
20. a (Similar kinds of intermolecular forces, in particular, hydrogen bonding — Chapter 9.)
21. c (Sulfuric acid would remove the lab coat!)
22. c (For the solvent $P_1 = X_1 P_1^0$ and $X_1 = 2/3$; Raoult's Law.)
23. b ($P_{tot} = X_1 P_1^0 + X_2 P_2^0$)
24. d (Depends on composition of solute particle.)
25. d (π = MRT = 1 × 8.3 × 300 = 2500 kPa)
26. a (Most sensitive for very dilute solutions. Why would you expect the solution to be dilute?)
27. d (Nonvolatile solute lowers the vapor pressure and mp.)
28. d (Can you show that M and m become equal at very low concentration?)
29. a ($\Delta T_f = k_f \cdot \dfrac{g/GMM}{kg\ solvent}$; make sure units are consistent!)
30. a (The largest number of moles of solute are used.)
31. b (As solvent freezes out, the solution remaining becomes more concentrated.)
32. (a) X sugar = $\dfrac{12.0/342}{(12.0/342) + (200/18.0)} = \dfrac{0.0351}{11.1} = 0.003\ 16$

(b) m = $\dfrac{12.0/342}{0.200} = 0.175$

(c) V(solution) = $\dfrac{212\ g}{1.022\ g/cm^3} = 207\ cm^3$

M = $\dfrac{12.0/342}{0.207} = 0.170$

33. (a) $0.065 = 1.86\ m = \dfrac{1.86\ (0.946)}{(GMM)(0.150)}$; GMM = 180 g/mol
    (b) FM = 30.0; 180/30.0 = 6; $C_6H_{12}O_6$

34. $0.261 = M(8.31)(300)$

    $M = 1.05 \times 10^{-4} \dfrac{mol}{dm^3} = \dfrac{1.00\ g}{(GMM)(0.100\ dm^3)}$; GMM = 95 200 g/mol

*35. At equilibrium, the vapor pressures above the solutions are the same; each is $XP°$.

$$g\ H_2O\ (\text{sugar solution}) = 24.90 - 1.68 = 23.22$$
$$g\ H_2O\ (\text{unknown solution}) = 20.00 - 3.22 = 16.78$$
$$X_{H_2O}(\text{sugar solution}) = X_{H_2O}(\text{unknown solution})$$

$$\dfrac{23.22/18.02}{\dfrac{23.22}{18.02} + \dfrac{1.68}{342}} = \dfrac{16.78/18.02}{\dfrac{16.78}{18.02} + \dfrac{2.45}{MM}}$$

$$MM = 670$$

*36. $k_b = \Delta T_b/m = 2.4$

$$47.2 - 46.1 = 2.4 \times \dfrac{1.5/MM}{0.0125}$$

$$MM = 260$$

$$S_8$$

SELECTED READINGS

Colligative properties are correlated with vapor pressure; how is vp lowering explained? Consider:

Mysels, K. J., The Mechanism of Vapor-Pressure Lowering, *Journal of Chemical Education* (April 1955), p. 179.

Problem solving (concentrations, solution stoichiometry . . .) is illustrated in the manuals listed in the Preface.

Unusual solutions (from electrons to macromolecules) are described in:

Dye, J. L., The Solvated Electron, *Scientific American* (February 1967), pp. 76-83.
Feeney, R. E., A Biological Antifreeze, *American Scientist* (November–December 1974), pp. 712-719.

13
WATER, PURE AND OTHERWISE

QUESTIONS TO GUIDE YOUR STUDY

1. Where and in what physical state do you find water?
2. Is pure water, like many other pure substances, found only in the laboratory, if even there?
3. What's so special about water? For example, how do ΔH_{vap}, mp, and bp compare to those for other substances of similar molecular mass? How do its solvent properties compare to those of other liquids?
4. What geometry and bonding have been described for the water molecule? How does this structure help explain water's unique properties?
5. How do water solutions compare to nonaqueous solutions? Do the same laws of colligative properties and the same principles of solubility apply?
6. What solutes are found in naturally occurring water solutions? What are their sources? How do they affect the uses of the water?
7. How can you determine experimentally the concentration of a solute in water solution?
8. How can you prepare very pure water? How is drinking water normally prepared? (What processes are used in a municipal water plant?)
9. Do we have, as for mixtures of gases, a workable model for salt solutions?
10. Can you justify an extensive study of water solutions? (Of the remaining chapters in the text, half are devoted to reactions which occur in water solution.)

174 • 13–Water, Pure and Otherwise

11.

12.

YOU WILL NEED TO KNOW

Concepts

1. How to write and interpret chemical equations — Chapter 3.
2. The formulas and charges of several common ions — Chapter 8.
3. The kinds and relative strengths of forces that exist between molecules — Chapter 9.
4. The physical properties of water — Chapters 9, 11.
5. The laws for colligative properties of solutions; the molecular factors and external conditions determining solubility — Chapter 12.

Math

1. How to perform stoichiometric calculations (those based on chemical formulas and equations) — Chapter 3.
2. How to work problems involving concentration units — Chapter 12.

CHAPTER SUMMARY

The unique properties of water (e.g., high boiling point and heat of vaporization, density behavior near 0°C) are related to the ability of the H_2O molecule to form three-dimensional networks held together by hydrogen bonds. Such networks are known to exist in ice, accounting for its low density relative to liquid water. They are believed to persist, probably in modified form, in liquid water near the melting point. As the temperature is raised, there is a shift toward a more closely packed arrangement typical of normal liquids. Qualitatively at least, we can explain the unusual behavior of water near 0°C (maximum density at 4°C) in terms of an equilibrium between "flickering clusters" of hydrogen-bonded water molecules and a more closely packed structure.

So much is review (Chapters 9 and 11). Chapter 12 was largely devoted to the description of aqueous solutions of molecular solutes. The current chapter considers the perhaps more familiar role of water as a vehicle for electrolytes.

Despite a great deal of research in the area, the effect of electrolytes on water structure is still a matter of controversy. At low concentrations, below about 0.01 m, the colligative properties of a solution of a 1:1 electrolyte such as NaCl can be expressed by Equations 13.3–13.5. If the solute ions (e.g., Na^+, Cl^-) behaved as completely independent particles, i would be equal to 2. As we go to very, very dilute solutions (m → 0), where the ions are extremely far apart, i → 2. For other types of electrolytes, too, we always find that as m → 0, i approaches the number of moles of ions per mole of electrolyte. All of these equations break down at embarrassingly low concentrations; nevertheless, they can be applied to freezing and boiling points, and to osmotic pressure data in very dilute solutions, to obtain i. In this way, it is possible to deduce the way in which a species ionizes in water; we can demonstrate, for example, that H_3PO_4 forms H^+ and $H_2PO_4^-$ ions (i = 2), while H_2SO_4 gives $2 H^+ + SO_4^{2-}$ (i = 3).

Relatively small amounts of suspended or dissolved species in water can make it unsafe for drinking, unappealing for recreational purposes, or harmful in other ways to the environment. One measure of the extent of pollution of water is its BOD (biochemical oxygen demand), which tells us the concentration of oxidizable species in water. Consider, for example, an organic pollutant: in its tendency to be oxidized, even perhaps to the harmless products CO_2 and H_2O, the pollutant will consume (make a demand on) dissolved oxygen. Many inorganic contaminants as well can have undesirable effects. An example is the phosphate ion, present in many detergents, which perhaps contributes to the eutrophication of lakes. Compounds of certain of the heavy metals, notably mercury and lead, are known to be toxic even at low concentrations.

The reuse of water is an absolute necessity; it has been estimated that the available water resources in the U.S. are daily replenished by some $10^9 m^3$, while the current rate of water use has been estimated to be $10^9 m^3$ also! Among the processes commonly used in municipal water purification are sedimentation, coagulation, filtration, and disinfection. These are designed primarily to clarify water and destroy disease-carrying organisms. Yet another purification process is water softening: waters containing relatively high concentrations of Ca^{2+}, Mg^{2+}, or Fe^{3+} can be made more usable by passing through columns containing natural or synthetic zeolites, which replace the objectionable cations by Na^+ ions. Alternatively, the same result can be achieved by the lime-soda process in which Ca^{2+} ions are precipitated as $CaCO_3$.

BASIC SKILLS

This chapter is primarily descriptive; the concepts considered are largely qualitative in nature (e.g., water structure, water pollution, water purification). Most of this material lends itself to discussion and informed

176 • 13–Water, Pure and Otherwise

speculation rather than to formal problem solving. You will note that most of the problems at the end of the chapter require discussions or simple conversions. A few quantitative ideas are presented; the more important are considered below.

1. Given a series of salts at the same concentration in (dilute) water solution, compare them as to:
    a. electrical conductivity;
    b. freezing point lowering.

Compare the electrical conductivities and freezing point lowerings of 0.10 M solutions of KNO_3, $Mg(NO_3)_2$, and $Al(NO_3)_3$ to those of 0.10 M NaCl. _____

As indicated in Table 13.1 and the discussion preceding it, conductivity depends upon charge type. Since KNO_3, like NaCl, is a 1:1 salt, 0.10 M solutions of these two compounds should have about the same conductivities. $Mg(NO_3)_2$, a 2:1 salt, should have a conductivity about twice that of NaCl. Finally, the 3:1 salt $Al(NO_3)_3$ would be expected to have a conductivity three times that of NaCl at the same concentration.

Freezing point lowering depends primarily upon the number of moles of ions present in solution. For one mole of each salt, we have:

	No. moles cations	No. moles anions	Total no. moles
NaCl	1	1	2
KNO_3	1	1	2
$Mg(NO_3)_2$	1	2	3
$Al(NO_3)_3$	1	3	4

Consequently we would expect that, at equal concentrations, the freezing point lowering of KNO_3 would be about the same as that of NaCl; $Mg(NO_3)_2$ should have a freezing point lowering about 3/2 that of NaCl, and $Al(NO_3)_3$ twice that of NaCl.

It should be emphasized that these comparisons apply strictly only in very dilute solution, where interionic attractive forces are of relatively minor importance. In practice, they are fairly reliable at concentrations up to 0.10 mol/dm^3.

Skill 1(a) is required in Problems 13.5 and 13.24; Skill 1(b) is used in Problems 13.1, 13.6, and 13.25.

2. Use Equations 13.3–13.5 to estimate:
    a. ΔT_f, ΔT_b, or π for an electrolyte solution given or having calculated i and m;

b. the nature of ionization of an electrolyte, given ΔT_f, ΔT_b, or π at a known m.

What is the freezing point of a one molal NaCl solution if i at this concentration is 1.80? _____
Substituting directly into Equation 13.3:

$$\Delta T_f = 1.80(1.00)(1.86°C) = 3.35°C$$

Hence the freezing point of the solution must be $-3.35°C$. Note that if NaCl behaved ideally at this concentration (i = 2), ΔT_f would be somewhat larger, $3.72°C$.

Skill 2(b) is illustrated in Example 13.1. See also Problems 13.7 and 13.26, which are very similar to Example 13.1.

3. **Calculate the BOD of a water sample, given the concentration of oxidizable material present.**

This skill is illustrated in Example 13.2. Note that the only new concept involved here is the definition of BOD (mg O_2 required /dm³ of sample). Otherwise, the calculation is a simple exercise in mass relations in chemical reactions (Chapter 3).
Problem 13.13 is very similar to Example 13.2; in Problem 13.32, the calculation is reversed to find the concentration of CN^- present, given the BOD.

4. **Given the concentrations of Ca^{2+} and HCO_3^- in a water sample of known volume, calculate how much $Ca(OH)_2$ and Na_2CO_3 should be added to soften it.**

As indicated in the discussion in the text, the most economical way to soften water by the lime-soda process is to:
    a. add one mole of $Ca(OH)_2$ per two moles of HCO_3^-; this removes one mole of Ca^{2+} from the hard water;
    b. remove any Ca^{2+} remaining by adding Na_2CO_3 in a 1:1 mole ratio. This principle is applied in Example 13.3 and in Problems 13.3, 13.19, and 13.38.

SELF-TEST

True or False

1. Seawater might reasonably be described as being a 0.5 M ( ) NaCl solution.

2. The maximum density of water at 4° C can be explained by ( ) assuming that the molecular structure of ice does not disappear completely when it melts.

3. A 0.1 m solution of KCl should freeze at about $-0.19°C$. ( )

4. It is believed that the $(C_2H_5)_4N^+$ ion promotes the ( ) formation of ice-like clusters of water molecules. Consequently, one would expect the maximum density of solutions containing this ion to be above 4°C.

5. A high BOD suggests that a water supply is contaminated ( ) with organic wastes.

6. Branched-chain detergents are more readily broken down in ( ) nature than those with straight chains.

7. In the lime-soda process of water softening, Ca^{2+} ions in the ( ) hard water are replaced by Na^+ ions.

8. A zeolite column used to soften water could be regenerated ( ) by flushing with a concentrated solution of $CaCl_2$.

9. Salt water may someday be converted on a large scale to ( ) pure water by the process of osmosis.

10. The following 0.01 M aqueous solutions are already ( ) arranged in order of increasing electrical conductivity: LiCl, $MgCl_2$, $FeCl_3$, CH_2Cl_2.

Multiple Choice

11. The orientation of covalent and hydrogen bonds about an ( ) oxygen atom in ice is best described as
    (a) bent   (b) planar
    (c) tetrahedral   (d) hexagonal

12. The property of water that is most critical to the regulation ( ) of body temperature is
    (a) high density at 4°C   (b) high boiling point
    (c) high heat of vaporization   (d) high surface tension

13. Of the following substances, the one expected to be most ( )
soluble in H_2O is
   (a) $C_2H_5OH(l)$
   (b) C_2H_5-O-C_2H_5 (l)
   (c) $C_4H_9OH(l)$
   (d) naphthalene (s)

14. To prepare 250 cm³ of 6.0 M NH_3(aq), you might take ( )
100 cm³ of 15 M NH_3(aq) and add to it about
   (a) 250 cm³ H_2O
   (b) 150 cm³ H_2O
   (c) 0.15 cm³ H_2O
   (d) nothing

15. At low concentrations, i for $Fe(NO_3)_3$ approaches ( )
   (a) 1
   (b) 2
   (c) 3
   (d) 4

16. For a 0.1 m solution of KNO_3, i would probably be ( )
   (a) about 1
   (b) slightly less than 2
   (c) slightly greater than 2
   (d) about 5

17. To explain the observation that i may be less than 2 at any ( )
finite concentration for aqueous solutions of a strong 1-1 or 2-2
electrolyte, it has been proposed that
   (a) molecules and ion pairs exist in solution
   (b) ions do not act independently of one another
   (c) water undergoes structural changes
   (d) all the above

18. A 0.010 m aqueous solution of which substance, listed ( )
below, would have the lowest solvent vapor pressure?
   (a) Br_2 (l)
   (b) CH_2Br_2 (l)
   (c) KBr(s)
   (d) $CaBr_2$(s)

19. Which one of the following 0.01 M water solutions would ( )
have the lowest freezing point?
   (a) CH_3OH
   (b) NaCl
   (c) $MgCl_2$
   (d) $MgSO_4$

20. A 0.010 M solution of $MgSO_4$(aq) should contain ( )
   (a) only particles of molecular or atomic dimensions
   (b) somewhat less than 0.020 M concentration of solute particles
   (c) few $MgSO_4$ molecules, and certainly no Mg or SO_2 molecules
   (d) all the above

21. The freezing point of a 0.0100 m solution of AB(aq) is ( )
−0.0186°C. What percentage of the AB molecules are undissociated?
   (a) 100
   (b) 99
   (c) 1.0
   (d) 0

180 • 13–Water, Pure and Otherwise

22. An anion which apparently promotes the growth of algae in water is ( )
   (a) Cl^-
   (b) SO_4^{2-}
   (c) PO_4^{3-}
   (d) CN^-

23. The principal advantage of an insecticide like Sevin over DDT is that it ( )
   (a) is less expensive
   (b) is less toxic to humans
   (c) remains effective longer
   (d) breaks down more quickly

24. Mercury-containing organic compounds resemble DDT in that they ( )
   (a) have similar toxicities
   (b) usually arise from the same source
   (c) accumulate in the food chain
   (d) have similar molecular structures

25. Which one of the following processes generally does not remove suspended matter from water? ( )
   (a) sedimentation
   (b) coagulation
   (c) filtration
   (d) chlorination

26. Residual Cl_2 in drinking water may be present at concentrations of one part per million. This corresponds to a molal concentration of about ( )
   (a) 1×10^{-6}
   (b) 1×10^{-5}
   (c) 1×10^{-3}
   (d) 1

27. One way to remove the unpleasant taste of chlorinated water is to ( )
   (a) filter it
   (b) use activated charcoal
   (c) add NaOH
   (d) soften it

28. A certain water supply contains Ca^{2+} and HCO_3^- in a 1:2 mole ratio. To soften this water by the lime-soda process, one should add ( )
   (a) only lime
   (b) only soda
   (c) both lime and soda
   (d) CO_2

29. In reverse osmosis, the relationship between the applied pressure P and the osmotic pressure π would most likely be ( )
   (a) $P < \pi$
   (b) $P = \pi$
   (c) $P > \pi$
   (d) $P \gg \pi$

30. Which one of the following substances would increase the BOD of a water supply? ( )
   (a) CO_2
   (b) O_3
   (c) C_2H_5OH
   (d) H_2O

31. The natural decomposition of organic matter in the absence ( ) of oxygen is likely to result in the formation of all the following except
  (a) H_2S
  (b) CH_4
  (c) NH_3
  (d) CO_2

32. Which of the following is generally *not* considered a water ( ) pollutant?
  (a) Na^+
  (b) Fe^{3+}
  (c) Ca^{2+}
  (d) all may be pollutants

33. What happens to the amount of dissolved oxygen in a ( ) stream as the temperature rises?
  (a) it increases
  (b) it decreases
  (c) it remains unchanged
  (d) cannot say

Problems

34. A certain substance partially ionizes in water as follows:

$$HX(aq) \rightleftarrows H^+(aq) + X^-(aq)$$

The freezing point of a 0.100 m solution of HX is $-0.200°C$.
  (a) Calculate the value of i in the equation $\Delta T_f = i(1.86°C)m$
  (b) Determine the % dissociation of HX in 0.100 m solution.

35. Consider the reaction between urea and oxygen:

$$2\ CO(NH_2)_2(aq) + 3\ O_2(g) \rightarrow 2\ N_2(g) + 2\ CO_2(g) + 4\ H_2O$$

A certain water supply contains one kilogram of urea in 10^4 dm³ of water.
  (a) How many grams of O_2 are required to react with the urea?

    (AM C = 12.0, N = 14.0, H = 1.0)

  (b) What is the BOD of the water?

36. How many moles of $Ca(OH)_2$ and Na_2CO_3 should be added to soften 2.0×10^5 dm³ of water if the concentrations of Ca^{2+} and HCO_3^- are, respectively,
  (a) 1.2×10^{-4} M and 2.4×10^{-4} M?
  (b) 1.2×10^{-4} M and 1.6×10^{-4} M?

*37. Read and summarize for oral presentation the work of Arrhenius on electrolyte solutions as reported in:

Arrhenius, S., "Über die Dissociation der in Wasser gelösten Stoffe," *Zeitschrift für physikalische Chemie*, Volume 1, pp. 631–648 (1887).

*38. Arrange for your class to visit a local waste-water treatment facility. Prepare a list of questions to ask.

SELF-TEST ANSWERS

1. T (At least in terms of its electrolytes.)
2. T (See Chapters 9, 11.)
3. F (Closer to $-0.37°C$; limiting value of i = 2.)
4. T (More clusters would exist at higher T than in pure H_2O.)
5. T (Usually, although inorganic solutes may also consume O_2.)
6. F (Straight chains more biodegradable.)
7. T
8. F (Use NaCl; Na^+ is to be exchanged for any Ca^{2+} in hard water.)
9. F (Reverse osmosis.)
10. F (CH_2Cl_2, a nonelectrolyte, should be first.)
11. c (Chapter 11.)
12. c (Would you expect specific heat to be important too?)
13. a (Of the two alcohols, both hydrogen-bonded, this has the smaller paraffin group. See Chapter 12 on solubility principles.)
14. b (Since 100 cm³ of 15 M NH_3 contains the needed number of moles NH_3, add enough H_2O to give 250 cm³ — Chapter 12.)
15. d (Four ions per mole: Fe^{3+} + 3 NO_3^-. For kinds of ions to expect, see Chapter 8.)
16. b (Attractive forces between the K^+ and NO_3^- ions effectively reduce their number.)
17. d
18. d (Largest number of solute particles. What would be the equation for vapor pressure lowering for electrolytes? For nonelectrolytes, see Chapter 12.)
19. c (Again, largest number of solute particles.)
20. d (You certainly should expect Mg^{2+} and SO_4^{2-} as structural units — see Chapters 8, 9.)
21. a (i = 1; each mole of AB gives one mole of solute particles.)
22. c
23. d
24. c
25. d (Cl_2 is used primarily for disinfection.)
26. b $\left(\dfrac{10^{-3} \text{ g } Cl_2/(71 \text{ g/mol})}{\text{kg } H_2O} \right)$
27. b
28. a (Add one mole of $Ca(OH)_2$ for every two moles of HCO_3^-.)
29. c (If $P \gg \pi$, the membrane might break.)
30. c (The carbon compound can consume O_2 to give CO_2.)
31. d (Organic carbon would be converted to CH_4 rather than CO_2: you would expect little, if any, combination with oxygen.)

32. d (Depends on intended use.)
33. b (One of the more drastic effects of thermal pollution.)
34. (a) i = 0.200/(1.86)(0.100) = 1.08
    (b) 8%
35. (a) 1000 g urea $\times \dfrac{96.0 \text{ g } O_2}{120 \text{ g urea}} = 800$ g O_2

    (b) 8.00×10^5 mg $O_2/10^4$ dm^3 = 80.0 mg O_2/dm^3
36. (a) $1.2 \times 10^{-4} \dfrac{\text{mol Ca(OH)}_2}{\text{dm}^3} \times 2.0 \times 10^5$ dm^3 = 24 mol Ca(OH)$_2$

    (b) $0.80 \times 10^{-4} \dfrac{\text{mol Ca(OH)}_2}{\text{dm}^3} \times 2.0 \times 10^5$ dm^3 = 16 mol Ca(OH)$_2$

    $0.40 \times 10^{-4} \dfrac{\text{mol Na}_2\text{CO}_3}{\text{dm}^3} \times 2.0 \times 10^5$ dm^3 = 8 mol Na$_2$CO$_3$

*37. An interlibrary loan may be required even to locate this article. A good dictionary of scientific German may be needed, for example:

   deVries, L., *German-English Science Dictionary,* New York, McGraw-Hill, 1959.

   What aspects of the Arrhenius theory have had to be modified?

*38. A few suggested questions:

   What is the average BOD of wastes entering the facility? Of wastes leaving the facility?

   How many people benefit directly from the service? What is the per capita cost of the treatment?

   What chemicals are used in each stage of purification?

   What contaminants are recoverable?

   What changes are likely in the next ten or twenty-five years?

SELECTED READINGS

Environmental chemistry is the subject of the journal Environmental Science and Technology *as well as the following (some of it dated, but some good chemistry discussed as well):*

Cleaning Our Environment: The Chemical Basis for Action, Washington, D. C., American Chemical Society, 1969.

Goldwater, L. J., Mercury in the Environment, *Scientific American* (May 1971), pp. 15–21.

Stoker, H. S., *Environmental Chemistry: Air and Water Pollution,* Glenview, Ill., Scott, Foresman, 1972.

13–Water, Pure and Otherwise

Water and natural water systems are discussed in:

Frank, H. S., The Structure of Ordinary Water, *Science* (August 14, 1970), pp. 635–641.

MacIntyre, F., The Top Millimeter of the Ocean, *Scientific American* (May 1974), pp. 62–77.

Probstein, R. F., Desalination, *American Scientist* (May-June 1973), pp. 280–293.

Runnels, L. K., Ice, *Scientific American* (December 1966), pp. 118–126.

14

SPONTANEITY OF REACTION; ΔG AND ΔS

QUESTIONS TO GUIDE YOUR STUDY

1. What do you mean by reaction *spontaneity*? (What special sense of the word is implied here?)
2. Can you decide whether or not a particular reaction will occur without even trying to carry out the reaction? (Recall being able to calculate ΔH in such a case.)
3. What physical meaning do you associate with ΔG and ΔS (and ΔH)? What kinds of measurements do you make to determine their values?
4. How are ΔG and ΔS (and ΔH) related to the masses of substances taking part in a reaction? Does Hess's Law apply to ΔG and ΔS?
5. How are these quantities related to each other (and to ΔH)? How is each of these related to the properties of atoms and molecules, like bond energy and molecular structure?
6. How do ΔG and ΔS (and ΔH) depend on reaction conditions such as temperature and pressure? Can you predict the sign or magnitude of the effect of a change in conditions for ΔG, ΔS, and ΔH?
7. Can you predict the sign or relative size of ΔS for a given reaction? Likewise for ΔG?
8. What happens at the molecular level when the entropy of a system increases? When free energy decreases?
9. To what kinds of systems or reaction conditions can you apply the principles of spontaneity discussed in this chapter?

10. Are you now able to decide the conditions under which any given reaction may occur? (Of what significance or usefulness is all this to the nonchemist or to society?)

11.

12.

YOU WILL NEED TO KNOW

Concepts

1. How to interpret the sign of ΔH, and ΔH itself, in terms of bond energies — Chapter 4.

Math

1. How to work problems in stoichiometry — Chapter 3.
2. How to calculate ΔH for any reaction — Chapter 4.

CHAPTER SUMMARY

If you were to ask several nonscientists to define a "spontaneous" process, you might get a variety of answers. One response might be: "a process taking place by itself without anyone having to work to bring it about." This statement resembles the definition of spontaneity presented in this chapter, where we say that (at constant temperature and pressure) *a spontaneous process is one which is capable of producing work.* Notice that it is the inherent capacity to do work that characterizes a spontaneous reaction. We say that the combustion of methane at 25°C and 1 atm is spontaneous because this reaction, if carried out in an appropriate machine, will produce energy in the form of work. The fact that under normal conditions the energy released when methane burns is dissipated as heat does not alter our conclusion.

The capacity of a reaction to produce work can be related to a fundamental property of substances known as free energy. The difference in free energy, ΔG, between products and reactants is a direct measure of the maximum amount of work that can be obtained from a reaction. Spontaneous reactions are those for which the products have a lower free energy than the reactants ($\Delta G < 0$). If the free energy of the products is greater than that of the reactants ($\Delta G > 0$), work must be done to make the reaction go and we say that it is nonspontaneous. If, perchance, the free energies of reactants and products are equal ($\Delta G = 0$), the reaction system is balanced on a knife edge; a tiny "push" will cause it to go in one direction or the other.

We can think of the free energy change as being made up of two components, according to the Gibbs-Helmholtz equation:

$$\Delta G = \Delta H - T\Delta S$$

The enthalpy change, ΔH, which we worked with in Chapter 4, represents the amount of heat absorbed or evolved when a reaction is carried out at constant pressure. If the bonds in the product molecules are stronger than those in the reactants, ΔH will be negative and the reaction will be exothermic. The Gibbs-Helmholtz equation tells us that a negative value of ΔH will tend to make ΔG negative as well. We conclude that, other things being equal, exothermic reactions will tend to be spontaneous.

The other quantity appearing in this equation, ΔS, represents the difference in entropy between products and reactants. Entropy is a measure of randomness. The entropy of a solution exceeds that of the pure components; gases have greater entropies than liquids or solids. We note from the Gibbs-Helmholtz equation that a positive value of ΔS will tend to make ΔG negative and hence contribute to spontaneity. This analysis confirms what experience tells us: order degenerates into chaos without any help from anyone.

The balance between these two factors depends upon the magnitude of the absolute temperature, T. At low temperatures, $T\Delta S$ will be small and the sign of ΔG will be that of ΔH. As temperature rises, the entropy factor plays a more important role and eventually dominates. If, as most often happens, ΔH and ΔS have the same sign, the direction in which a reaction proceeds spontaneously will reverse at some temperature. A simple example is the vaporization of water at 1 atm pressure (ΔH and ΔS both positive). Below 100°C, ΔH is greater than $T\Delta S$, ΔG is positive, and vaporization does not occur. At 100°C, $\Delta H = T\Delta S$, $\Delta G = 0$, and the system is at equilibrium. Above 100°C, ΔS predominates, $\Delta G < 0$, and water at atmospheric pressure boils spontaneously.

Of the three quantities in the Gibbs-Helmholtz equation: 1) ΔH is essentially independent of both temperature and pressure; 2) ΔS can be taken to be temperature independent, at least above room temperature, but is strongly dependent upon pressure for many reactions, in particular, those involving gases; and 3) ΔG is ordinarily dependent upon both temperature and pressure.

Throughout this chapter, we have confined our discussion to reactions taking place at 1 atm and hence have dealt with $\Delta G^{1\ atm}$. This quantity is often calculated from free energies of formation of compounds:

188 • 14—Spontaneity of Reaction; ΔG and ΔS

$$\Delta G^{1\ atm} = \Sigma \Delta G_f^1\ {}^{atm}\ products - \Sigma \Delta G_f^1\ {}^{atm}\ reactants$$

(Electrochemical measurements are generally the experimental source for these quantities, as discussed in Chapter 23, while measurements of heat capacity at different temperatures provide entropy values. How would you measure a ΔH?)

BASIC SKILLS

1. **Given a table of standard free energies of formation, calculate $\Delta G^{1\ atm}$ at 25°C for a reaction.**

The calculation here is entirely analogous to that used in Chapter 4 to obtain ΔH for a reaction from heats of formation. Since ΔG, unlike ΔH, varies considerably with both temperature and pressure, these two conditions must be specified; i.e., we calculate "$\Delta G^{1\ atm}$ at 25°C."

This skill is illustrated in Example 14.1 and applied directly in Problems 14.1, 14.11, and 14.27. In the example and in each of these problems, the significance of the sign of ΔG is emphasized. ΔG is negative for a spontaneous reaction and positive for a nonspontaneous reaction.

Several other problems at the end of Chapter 14 involve the use of this skill in combination with others. See, for example, Problems 14.17, 14.33 and, somewhat less obvious, 14.18–14.20 and 14.34–14.36. In general, whenever you are asked to calculate ΔS for a reaction or obtain ΔG as a function of temperature, you must first obtain ΔG at a fixed temperature, usually 298 K (Table 14.2), and ΔH (Table 4.1, Chapter 4).

2. **Apply the "laws of thermochemistry" (p. 338 of the text) to calculations involving ΔG.**

You will note that these laws are precisely the same as those discussed earlier in connection with ΔH in Chapter 4. See the examples in the body of the text and Problems 14.1d, 14.8b, 14.9, 14.24b, and 14.25. The following example may also be helpful.

Given that $2\ Sn(s) + O_2(g) \rightarrow 2\ SnO(s)$; $\Delta G = -515$ kJ

$2\ SnO(s) + O_2(g) \rightarrow 2\ SnO_2(s)$; $\Delta G = -525$ kJ

Calculate ΔG for $SnO_2(s) \rightarrow Sn(s) + O_2(g)$ _____
We start by adding the two equations given to obtain (upon cancelling the two moles of SnO on both sides):

$$2 \text{ Sn(s)} + 2 \text{ O}_2\text{(g)} \rightarrow 2 \text{ SnO}_2\text{(s)}; \Delta G = -1040 \text{ kJ}$$

Dividing by 2 we obtain:

$$\text{Sn(s)} + \text{O}_2\text{(g)} \rightarrow \text{SnO}_2\text{(s)}; \Delta G = -520 \text{ kJ}$$

Finally, reversing the reaction:

$$\text{SnO}_2\text{(s)} \rightarrow \text{Sn(s)} + \text{O}_2\text{(g)}; \Delta G = +520 \text{ kJ}$$

---

3. **Estimate the sign of ΔS for certain physical and chemical changes.**

The general principle here is that in going from a more ordered to a less ordered state, entropy increases. Examples of such processes are fusion, vaporization, and the formation of a solution. For chemical reactions, we can be sure that ΔS will be positive if there is an increase in the number of moles of gas. For reactions which do not involve gases, ΔS is usually positive if there is a large increase in the number of moles.

Predict the sign of ΔS for each of the following processes.

$\text{H}_2\text{O(l)} \rightarrow \text{H}_2\text{O (g)}$ _____

$\text{H}_2\text{O(l)} + \text{CH}_3\text{OH(l)} \rightarrow$ solution _____

$2 \text{ H}_2\text{(g)} + \text{O}_2\text{(g)} \rightarrow 2 \text{ H}_2\text{O(g)}$ _____

For the first two processes, we would expect ΔS to be positive since we are going from a more ordered state (pure liquid) to a less ordered state (gas or solution). For the chemical reaction, ΔS should be negative, since there is a decrease in the number of moles of gas (3 mol gas → 2 mol gas).

See Problems 14.3, 14.14, and 14.30, which include applications of this principle to areas outside the natural sciences.

4. **Given the heat of vaporization (heat of fusion) and normal boiling point (freezing point) of a substance, calculate $\Delta S^{1 \text{ atm}}$ for the process.**

This calculation is illustrated in Example 14.2. Note that in order to obtain $\Delta S^{1 \text{ atm}}$, we must use the normal boiling point or freezing point, i.e., the temperature at which the two phases are at equilibrium at 1 atm pressure. Only at that temperature is $\Delta G^{1 \text{ atm}} = 0$ and hence $\Delta S^{1 \text{ atm}} = \Delta H/T$.

190 • 14–Spontaneity of Reaction; ΔG and ΔS

Problems 14.2, 14.13, and 14.29 apply this skill.

5. Given or having calculated ΔH and $\Delta G^{1\ atm}$ at 25°C, use the Gibbs-Helmholtz equation (Equation 14.12) to obtain:
   a. $\Delta S^{1\ atm}$: $\Delta S^{1\ atm} = (\Delta H - \Delta G^{1\ atm}$ at 298K$)/298K$;
   b. $\Delta G^{1\ atm}$ at any temperature T: $\Delta G^{1\ atm} = \Delta H - T\Delta S^{1\ atm}$;
   c. the temperature at which the reaction is at equilibrium at one atmosphere pressure: $T = \Delta H/\Delta S^{1\ atm}$.

These several applications of the Gibbs-Helmholtz equation are illustrated in Examples 14.3 and 14.4. The example below illustrates the combined applications for a single reaction.

For a certain reaction, $\Delta H = +19.0$ kJ and $\Delta G^{1\ atm}$ at 25°C = +15.0 kJ. What is:
  a. $\Delta S^{1\ atm}$?
  b. $\Delta G^{1\ atm}$ at 1000K?
  c. T at which the reaction is at equilibrium at 1 atm pressure?

Applying the above equations:
$\Delta S^{1\ atm} = (\Delta H - \Delta G^{1\ atm}$ at 298K$)/T = (19.0$ kJ $- 15.0$ kJ$/298$K
$= 0.013$ kJ/K
$\Delta G^{1\ atm}$ at 1000K $= \Delta H - 1000(\Delta S^{1\ atm}) = 19.0$ kJ $- 13.0$ kJ
$= +6.0$ kJ

$$T = \frac{19.0 \text{ kJ}}{0.013 \text{ kJ/K}} = 1500\text{K}$$

Note that the quantities required for this calculation (ΔH and $\Delta G^{1\ atm}$ at 25°C) are ordinarily obtained from tables of heats and free energies of formation. It may also be worth pointing out that we do not really need to know $\Delta G^{1\ atm}$ at 25°C to obtain $\Delta S^{1\ atm}$. If, for example, we were given $\Delta G^{1\ atm}$ at 500K, we could obtain $\Delta S^{1\ atm}$ as:

$$\Delta S^{1\ atm} = (\Delta H - \Delta G^{1\ atm} \text{ at } 500\text{K})/500\text{K}$$

SELF-TEST

True or False

1. At 25°C and one atmosphere pressure, the reaction between ( ) N_2 and O_2 to form NO(g) is spontaneous.

2. For the decomposition of water to the elements at 25°C ( )
and 1 atm, ΔG = 237 kJ. This means that at least 237 kJ of work has
to be done to make this reaction go.

3. If ΔG for a reaction is positive, it is impossible to carry out ( )
the reaction unless either the temperature or the pressure is changed.

4. "Work," as the term is used in this chapter, includes radiant ( )
energy.

5. All exothermic reactions become spontaneous as we ( )
approach absolute zero.

6. Free energies of formation of compounds are positive ( )
numbers in most cases.

7. For the process $CO_2(s) \rightarrow CO_2(g)$, you should expect the ( )
signs of ΔH and ΔS to be the same.

8. For this same process, ΔS should decrease as pressure is ( )
increased.

9. For the reaction $PCl_5(g) \rightarrow PCl_3(g) + Cl_2(g)$, ΔS is positive. ( )

10. If ΔH and ΔS for a reaction are both negative, we expect ΔG ( )
to be negative at all temperatures.

Multiple Choice

11. For the combustion of methane at 25°C and 1 atm ( )

$$CH_4(g) + 2\,O_2(g) \rightarrow CO_2(g) + 2\,H_2O(l),$$

which one of the following statements is *not* true?
    (a) the reaction is exothermic
    (b) the reaction is spontaneous
    (c) $\Delta G < 0$
    (d) work has to be done to make the reaction go

12. Which one of the following quantities is independent of ( )
pressure?
    (a) ΔG (b) ΔH
    (c) ΔS (d) all of these

13. The free energy of formation of AgCl is −110 kJ/mol. ΔG ( )
for the reaction $2\,AgCl(s) \rightarrow 2\,Ag(s) + Cl_2(g)$ is
    (a) −220 kJ   (b) −110 kJ
    (c) +110 kJ   (d) +220 kJ

14–Spontaneity of Reaction; ΔG and ΔS

14. Which one of the following statements best describes the ( )
relationship between ΔG and temperature?
    (a) ΔG is independent of T
    (b) ΔG varies with T
    (c) ΔG is a linear function of T
    (d) ΔG usually decreases as T decreases

15. Which of the following would you expect to have the largest ( )
entropy per mole?
    (a) Li(g)   (b) Li(s)
    (c) $LiCl \cdot H_2O(s)$ (d) they should be the same

16. The heat of fusion of benzene is 10.7 kJ/mol. Its melting ( )
point is 5°C. ΔS for the melting of benzene, in J/(mol·K), is
about
    (a) 2.1   (b) 10.9
    (c) 38.5   (d) 54.4

17. The heat of vaporization of benzene is 31 kJ/mol; the ( )
normal boiling point is 80°C. Using this data, ΔS for the vaporization
of benzene *at 1 atm and 25°C* in kJ/(mol·K) is
    (a) 31/353   (b) 31/298
    (c) some other number   (d) impossible to calculate

18. For which of the following would you expect the magni- ( )
tude of ΔS to be the largest?
    (a) $C(s) + O_2(g) \to CO_2(g)$
    (b) $2 SO_2(g) + O_2(g) \to 2 SO_3(g)$
    (c) $CaSO_4(s) + 2 H_2O(l) \to CaSO_4 \cdot 2 H_2O(s)$
    (d) $3 H_2(g) + N_2(g) \to 2 NH_3(g)$

19. For a certain reaction, ΔH is 20 kJ and ΔG at 25°C is 14 kJ. ( )
ΔS in kJ/(mol·K) is approximately
    (a) +6   (b) +0.02
    (c) −0.02   (d) −6

20. Vaporization is an example of a process for which ( )
    (a) ΔH, ΔS, and ΔG are positive at all temperatures
    (b) ΔH and ΔS are positive
    (c) ΔG is negative at low T, positive at high T
    (d) ΔH is strongly pressure dependent

21. For a certain reaction, ΔH = +2.5 kJ, $\Delta S^{1\ atm}$ = 0.010 ( )
kJ/(mol·K). This reaction will be at equilibrium at 1 atm at about
    (a) 0.25°C   (b) 25°C
    (c) −23°C   (d) cannot tell

22. The free energy of formation of CO is −137 kJ/mol at 25°C; ( )
its heat of formation is −111 kJ/mol. As the temperature is
increased, ΔG_f will

(a) remain unchanged   (b) change sign
(c) become less negative (d) become more negative

23. When aniline dissolves in hexane, ΔS is positive. Why then is ( ) aniline only slightly soluble in hexane at room temperature? Choose the best thermodynamic explanation.
    (a) an equilibrium is reached, at which point, ΔG = 0
    (b) ΔH is positive
    (c) the intermolecular forces in the two liquids are very similar
    (d) the formation of a solution is always a spontaneous process

24. When the city of New York burns coal in an electrical ( ) generating plant, the entropy of the universe
    (a) increases   (b) decreases
    (c) is not changed (d) is zero

25. An emission control device in Ray's 1973 Volvo adsorbs ( ) gasoline vapors onto solid activated carbon. During the spontaneous process of adsorption
    (a) the carbon becomes warm
    (b) the carbon becomes cool
    (c) there is no change in T
    (d) the T may change, but one cannot say how

26. The reaction $CaO(s) + H_2O(l) \rightarrow Ca(OH)_2(s)$ is spontaneous ( ) at 25°C; the reverse reaction becomes spontaneous at high temperature. This means that
    (a) ΔH is +, ΔS is +   (b) ΔH is +, ΔS is −
    (c) ΔH is −, ΔS is −   (d) ΔH is −, ΔS is +

27. Each of the following reactions is nonspontaneous at 25°C ( ) and 1 atm. Which should become spontaneous at high temperature?
    (a) $CaCO_3(s) \rightarrow CaO(s) + CO_2(g)$
    (b) $CuCl_2(s) \rightarrow CuCl(s) + ½ Cl_2(g)$
    (c) $Fe_2O_3(s) + 1.5 C(s) \rightarrow 2 Fe(s) + 1.5 CO_2(g)$
    (d) all of them

28. The solubility of NaCl in water at 25°C is about 6 ( ) mol/dm³. Suppose you add 1 mol of NaCl to 1 dm³ of water. For the reaction $NaCl(s) + H_2O \rightarrow$ salt solution
    (a) ΔG > 0, ΔS > 0   (b) ΔG < 0, ΔS > 0
    (c) ΔG > 0, ΔS < 0   (d) ΔG < 0, ΔS < 0

29. For a certain reaction, $\Delta G^{1\ atm}$ is known at two different ( ) temperatures. From this information alone, one can calculate for the reaction

194 • 14—Spontaneity of Reaction; ΔG and ΔS

(a) ΔH
(b) ΔS¹ atm
(c) both ΔH and ΔS¹ atm
(d) neither ΔH nor ΔS¹ atm

Problems

30. Given the following information for the reaction

$$3\ Fe_2O_3(s) \rightarrow 2\ Fe_3O_4(s) + \tfrac{1}{2}\ O_2(g)$$

	ΔH_f (kJ/mol)	ΔG_f¹ atm (at 25°C, kJ/mol)
$Fe_3O_4(s)$	−1121	−1014
$Fe_2O_3(s)$	− 822	− 741
$H_2O(l)$	− 286	− 237

Calculate

(a) ΔG when one mole of Fe_2O_3 reacts at 25°C and 1 atm.
(b) ΔS when one mole of Fe_3O_4 is produced by the above reaction at 1 atm.

31. For a certain reaction, ΔH = −52.0 kJ. At 400K, ΔG for this reaction is −36.4 kJ. Determine

(a) ΔS for this reaction, in kJ/K.
(b) ΔG at 1000K.
(c) ΔG at 0K.

32. For the reaction $Cu_2O(s) + \tfrac{1}{2}\ O_2(g) \rightarrow 2\ CuO(s)$, the free energy change at 1 atm is −144 kJ at 0K and −108 kJ at 300 K. Calculate

(a) ΔH
(b) ΔS

*33. In the combustion of carbon and many of its compounds at or near room temperature and atmospheric pressure, one usually observes the formation of CO_2. Show by calculation why one expects CO_2 and not CO as the usual combustion product. (You might wish to choose a particular compound, or carbon itself, or perhaps try to relate CO and CO_2 directly.) Try to give a molecular level interpretation to your answer. Would you expect CO to be the major product at higher temperatures? Again, justify your answer by means of a calculation.

*34. Certain metals, when heated with excess $Cl_2(g)$, form two or more different chlorides, depending on the temperature. Invariably, it is the chloride containing the *lowest* percentage of chlorine which is formed at the highest temperatures (e.g., MCl_2 forms rather than MCl_3). Choose any metal that forms at least two chlorides and illustrate by appropriate calculation the validity of the above statement. Convenient tabulations of thermodynamic data may be found in the handbooks listed in the Preface to this guide.

SELF-TEST ANSWERS

1. **F** (This answer is suggested by the coexistence of N_2 and O_2, with very little NO, in the atmosphere, and confirmed by the fact that ΔG_f of NO is positive.)
2. **T** (Of course, *per mole* of H_2O.)
3. **F** (See Question 2; nonspontaneous does not mean impossible.)
4. **T** (Radiant energy brings about the nonspontaneous process of photosynthesis.)
5. **T** (ΔH determines the sign of ΔG at low T.)
6. **F** (Most compounds are stable relative to the elements at 25°C, 1 atm.)
7. **T** ($\Delta S > 0$ since Δn gas > 0; $\Delta H > 0$ for sublimation, Chapter 11.)
8. **T** (Entropy of gas will decrease, while that of solid will remain about the same; their *difference* decreases.)
9. **T** (Δn gas > 0.)
10. **F** (Positive at high T.)
11. **d** (Even if a small push is required to get it going, the reaction can produce work overall.)
12. **b**
13. **d** (ΔG_f applies to $Ag(s) + \frac{1}{2} Cl_2(g) \rightarrow AgCl(s)$. Use the thermochemical laws.)
14. **c** ($\Delta G = \Delta H - T\Delta S$, where ΔH and ΔS are constants, independent of T.)
15. **a** (Gaseous.)
16. **c** ($\Delta G = 0$ at equilibrium T, so $\Delta S = \Delta H/T$.)
17. **a** ($\Delta S_{vap}^{1\ atm} = \dfrac{\Delta H_{vap}}{T_b}$; ΔS changes very little with T.)
18. **d** (Δn gas is largest.)
19. **b** (($\Delta H - \Delta G)/T = 6/300$.)
20. **b**
21. **c** ($T = \Delta H/\Delta S = 250K$.)
22. **d** (ΔS is positive.)
23. **b** (The two driving forces are opposed.)
24. **a** (Why?)
25. **a** (Figure this on the basis of the signs of ΔG and ΔS.)
26. **c** (You expect $\Delta H < 0$ since it gives ΔG its sign at low T; ΔS determines sign of ΔG at high T.)
27. **d** (Look for $\Delta S > 0$ to make reactions spontaneous at high T.)

196 • 14—Spontaneity of Reaction; ΔG and ΔS

28. b (For any amount up to 6 mol/dm^3, $\Delta G < 0$ for solution formation; ΔS will always be positive for mixing.)

29. c (As well as ΔG at any other T. The data provides two equations in two unknowns, of the type $\Delta G = \Delta H - T\Delta S$.)

30. (a) $Fe_2O_2(s) \rightarrow \frac{2}{3} Fe_3O_4(s) + \frac{1}{6} O_2(g)$
    $\Delta G = \frac{2}{3}(-1014 \text{ kJ}) + 741 \text{ kJ} = +65 \text{ kJ}$

    (b) $\frac{3}{2} Fe_2O_3(s) \rightarrow Fe_3O_4(s) + \frac{1}{4} O_2(g)$
    ΔG at $25°C = -1014 \text{ kJ} + 1.5(741 \text{ kJ}) = +97 \text{ kJ}$
    $\Delta H = -1121 \text{ kJ} + 1.5(822 \text{ kJ}) = +112 \text{ kJ}$
    $\Delta S = 15 \text{ kJ}/298 \text{ K} = 0.050 \text{ kJ/K}$

31. (a) $\Delta S = \dfrac{-52.0 \text{ kJ} + 36.4 \text{ kJ}}{400 \text{ K}} = -0.0390 \text{ kJ/K}$

    (b) $\Delta G = -52.0 \text{ kJ} - 1000 \text{ K}(-0.0390 \text{ kJ/K}) = -13.0 \text{ kJ}$

    (c) -52.0 kJ

32. (a) -144 kJ

    (b) $\dfrac{-144 \text{ kJ} + 108 \text{ kJ}}{300 \text{K}} = -0.12 \text{ kJ/K}$

*33. Suppose CO were formed; the further reaction $CO(g) + \frac{1}{2} O_2(g) \rightarrow CO_2(g)$ would occur since $\Delta G^{1 \text{ atm}} = -257.1 \text{ kJ}$ at $25°C$. One can show that as T increases, ΔG becomes less negative, and reaction is less likely to occur. (ΔS favors the reverse reaction.) See Section 17.3 of the text.

*34. Consider $CrCl_3(s) \rightarrow CrCl_2(s) + \frac{1}{2} Cl_2(g)$
    $\Delta H = \Delta H_f \; CrCl_2 - \Delta H_f \; CrCl_3 = -396 \text{ kJ} + 563 \text{ kJ} = +167 \text{ kJ}$
    $\Delta G = \Delta G_f^{1 \text{ atm}} CrCl_2 - \Delta G_f^{1 \text{ atm}} CrCl_3 = +137 \text{ kJ}$ at $25°C$, 1 atm
    $\Delta S = 30 \text{ kJ}/298\text{K} = 0.10 \text{ kJ/K}$
    $\Delta G_T = 167 \text{ kJ} - 0.10 \text{ T}$; reaction reverses at about 1700 K
    For all such reactions, both ΔH and ΔS are positive, as in the above example.

SELECTED READINGS

Alternative discussions of thermodynamics, from a highly organized outline of working principles to a detailed statistical approach (looking at molecular activity), are presented in the books by Mahan and Pimentel (Chapter 4) and:

Bent, H. A., *The Second Law*, New York, Oxford University Press, 1965.
Campbell, J. A., *Why Do Chemical Reactions Occur?*, Englewood Cliffs, N. J., Prentice-Hall, 1965.
MacWood, G. E., How Can You Tell Whether a Reaction Will Occur?, *Journal of Chemical Education* (July 1961), pp. 334–337.
Nash, L. K., *Chemthermo*, Reading, Mass., Addison-Wesley, 1972.

Porter, G., The Laws of Disorder, *Chemistry* (May 1968), pp. 23–25.
Sanderson, R. T., Principles of Chemical Reaction, *Journal of Chemical Education* (January 1964), pp. 13–22.

Other topics related to those of this chapter are found in:

Bent, H. A., Haste Makes Waste, Pollution and Entropy, *Chemistry* (October 1971), pp. 6–15.
Layzer, D., The Arrow of Time, *Scientific American* (December 1975), pp. 56–69.
Smith, W. L., Thermodynamics, Folk Culture, and Poetry, *Journal of Chemical Education* (February 1975), pp. 97–98.

15
CHEMICAL EQUILIBRIUM IN GASEOUS SYSTEMS

QUESTIONS TO GUIDE YOUR STUDY

1. How would you show experimentally that a given system is in a state of equilibrium?

2. If you could watch the molecules in an equilibrium system, what would you expect to see?

3. The preceding chapter established a thermodynamic criterion for recognizing equilibrium. What is it?

4. How do you interpret equilibrium in terms of the "driving forces" behind changes in chemical systems, ΔH and ΔS?

5. How can you predict the conditions under which equilibrium may exist? For example, at what temperatures and pressures are ice and water in equilibrium? Or, hydrogen, oxygen, and water at 1 atm?

6. Can you predict the effect of changes in conditions on a system already at equilibrium? (Recall that you have been able to predict whether or not a given reaction may spontaneously occur; whether or not a phase change may occur.)

7. Is there a general approach to describing equilibrium systems? (Chapter 11 dealt with phase equilibria; Chapters 18–23 deal with several types of equilibria in water solution.)

8. Can you describe a common example of a reaction that only partially converts reactants to products?

9. If many reactions do not go to completion, what does it mean about how you are to interpret chemical equations?

200 • 15–Chemical Equilibrium in Gaseous Systems

10. How might you show experimentally that a given reaction is reversible?

11.

12.

YOU WILL NEED TO KNOW

Concepts

1. How to interpret the signs of ΔH and ΔG – Chapters 4, 14.
2. The general nature of an equilibrium system; the effects of changes in conditions; the interpretation in terms of molecular behavior – these topics are reviewed here and discussed in detail (for phase equilibria) in Chapter 11.

Math

1. How to solve quadratic equations, using the quadratic formula; how to calculate square roots (e.g., by use of slide rule or logs, if not by electronic calculator) – Readings listed in Preface.
(For the equation $ax^2 + bx + c = 0$,

$$x = \frac{-b \pm \sqrt{b^2 - 4ac}}{2a}$$

Note that here only one of the two possible solutions for x will make any physical sense.)
2. How to perform stoichiometric calculations – Chapter 3.
3. How to calculate ΔH and ΔG for any reaction – Chapters 4, 14.
4. How to work problems involving concentration units – Chapter 12.
5. How to calculate logs and antilogs – Appendix 4.

CHAPTER SUMMARY

We have seen (Chapters 11, 14) that a closed system will spontaneously undergo a change at constant temperature and pressure (i.e., a reaction will occur) if by so doing its free energy can decrease. That is, the free energy of a system moves toward a minimum value. What happens once the free energy is at this minimum value, even though, as is often the case, measurable amounts of reactants have not been consumed? Nothing that you can see! Reaction ceases; the system, unless disturbed, is in a state of *dynamic equilibrium*. (Dynamic, because our notion that molecules are always going about their business is supported by experiment.) This state of apparent rest can be considered the result of two driving forces striking a balance: the tendency of a system to move toward a state of minimum enthalpy; the tendency toward maximum entropy ($\Delta G = 0$ and so $\Delta H = T\Delta S$). Another interpretation of the equilibrium state considers that the rates of competing reactions have become equal; no net, observable change occurs as a result (Chapter 16).

The extent of a reaction for a system at equilibrium can be described by a certain ratio of concentrations. This ratio, the *equilibrium constant* K_c, is found to have a value which depends only on the temperature (much as the vapor pressure of a pure substance depends only on the temperature). Its value does not depend, for example, on the concentrations of reactants or products before equilibrium is reached, or on how the equilibrium state is approached. Of course, it does depend on the particular substances taking part in the reaction.

K_c has the same form for all equilibrium systems. For the general reaction of gaseous or dissolved species, A, B, C, and D: $aA + bB \rightleftharpoons cC + dD$

$$K_c = \frac{[C]^c [D]^d}{[A]^a [B]^b}$$

where the brackets refer to the concentration at equilibrium. This expression is commonly referred to as the Law of Chemical Equilibrium. Like all the laws we have encountered, it is an approximation to real behavior. The more nearly gases behave like ideal gases, the more nearly does this ratio have a constant value at a given temperature. In discussing equilibria in water solutions, we will see that the law holds very well only for very dilute solutions.

We can decide whether a given system is already in a state of equilibrium, or, if not, how it will approach equilibrium. We need only to compare the given concentrations to the ratio K_c at this same temperature. (Comparison to K_c allows us to predict the change, if any, that is spontaneous, just as we have seen whether or not the free energy could decrease.)

Also by comparison to K_c we can predict how a given change in conditions will affect a system already at equilibrium. The same prediction can be made qualitatively by use of *Le Chatelier's Principle:* When an equilibrium system is subjected to a change in conditions, a reaction will occur so as to minimize the effect of the change. (The system will again move toward a state of minimum free energy.)

202 • 15–Chemical Equilibrium in Gaseous Systems

Nothing yet can be said about how fast equilibrium is approached; neither ΔG nor K_c tells us anything about reaction rates. We will consider this topic in the following chapter.

BASIC SKILLS

1. Given the balanced equation for a reaction involving gases, write the corresponding expression for K_c.

The principles involved here are discussed in Section 15.2 of the text. Note particularly that terms for pure solids or liquids do not appear in the expression for K_c.

What is the expression for K_c for:

$$4 NH_3(g) + 7 O_2(g) \rightleftharpoons 4 NO_2(g) + 6 H_2O(l)?$$ _____

We note that the term for the gaseous product, NO_2, appears in the numerator, while those for the reactants, NH_3 and O_2, are in the denominator. No term for H_2O is included since it is a liquid. The exponents of the concentration terms are the coefficients in the balanced equation.

$$K_c = \frac{[NO_2]^4}{[NH_3]^4 [O_2]^7}$$

See Problems 15.1, 15.9, and 15.25. Note that in Problem 15.25d you must first write the balanced equation from the information given.

2. Given K_c for a particular equation, obtain K_c for other equations that could be written to represent the same equilibrium system.

This skill is discussed on p. 360 of the text. Note that in general:

a. *K_c for the reverse reaction is the reciprocal of the equilibrium constant for the forward reaction.* For example, if for the reaction

$$N_2(g) + O_2(g) \rightleftharpoons 2 NO(g); K_c = 0.10$$

then for the reverse reaction:

$$2\ NO(g) \rightleftharpoons N_2(g) + O_2(g); K_c = 1/0.10 = 10$$

b. *If the coefficients in the balanced equation are multiplied by a factor, x, the equilibrium constant is raised to the power, x.* For example:

if $\quad N_2(g) + O_2(g) \rightleftharpoons 2\ NO(g); K_c = 0.10$

then $\quad 2\ N_2(g) + 2\ O_2(g) \rightleftharpoons 4\ NO(g); K_c = (0.10)^2 = 0.010; x = 2$

and $\quad \frac{1}{2}N_2(g) + \frac{1}{2}O_2(g) \rightleftharpoons NO(g); K_c = (0.10)^{1/2} = 0.32; x = 1/2$

This skill is applied in Problems 15.10 and 15.26. It will be useful later on when we consider equilibria in aqueous solution, particularly in Chapters 19 and 20 (acid-base equilibria).

3. Apply the Rule of Multiple Equilibria to obtain K_c for a reaction.

This skill is discussed on p. 360 of the text; note particularly the example in the body of the text. The skill is applied in its simplest form in Problem 15.2. In Problems 15.11 and 15.27 and in the following example it must be combined with Skill 2, above.

Given: (1) $CuO(s) + H_2(g) \rightleftharpoons Cu(s) + H_2O(g); K_c = 2 \times 10^{15}$

(2) $H_2(g) + \frac{1}{2}O_2(g) \rightleftharpoons H_2O(g); K_c = 5 \times 10^{22}$

Calculate K_c for the reaction $CuO(s) \rightleftharpoons Cu(s) + \frac{1}{2}O_2(g)$ _____

The approach here is to combine equations (1) and (2) in such a way as to arrive at the desired equation. This can be done by reversing (2) and adding it to (1). Recalling that the equilibrium constant for the reverse reaction is the reciprocal of that for the forward reaction, we have

$CuO(s) + H_2(g) \rightleftharpoons Cu(s) + H_2O(g);\quad K_c = 2 \times 10^{15}$

$H_2O(g) \rightleftharpoons H_2(g) + \frac{1}{2}O_2(g)\quad;\quad K_c = \dfrac{1}{5 \times 10^{22}}$

$CuO(s) \rightleftharpoons Cu(s) + \frac{1}{2}O_2(g)\quad;\quad K_c = \dfrac{2 \times 10^{15}}{5 \times 10^{22}} = 4 \times 10^{-8}$

with the Rule of Multiple Equilibria applied to obtain the final equilibrium constant.

15–Chemical Equilibrium in Gaseous Systems

This skill, like Skill 2, will be very useful in later chapters.

4. For a given equation, calculate the numerical value of K_c knowing:
   a. the equilibrium concentrations of all species.

For the reaction $2\ NO(g) + O_2(g) \rightleftharpoons 2\ NO_2(g)$, the equilibrium concentrations of NO_2, NO, and O_2 are 1.0×10^{-2}, 1.0×10^{-3}, and 0.50×10^{-1} mol/dm^3 respectively. What is K_c? _____

All that is required here is to set up the expression for K_c and substitute the concentrations given:

$$K_c = \frac{[NO_2]^2}{[NO]^2[O_2]} = \frac{(1.0 \times 10^{-2})^2}{(1.0 \times 10^{-3})^2 (0.50 \times 10^{-1})} = 2.0 \times 10^3$$

This skill is applied in a straightforward way in Problems 15.12 and 15.28.

b. the original concentrations of all species and the equilibrium concentration of one species.

This skill is slightly more difficult to apply than 4(a). You may find it helpful to set up a table such as the one in the example below.

For the reaction $2\ HI(g) \rightleftharpoons H_2(g) + I_2(g)$, if we start with pure HI at a concentration of 1.20 mol/dm^3, that of H_2 at equilibrium is 0.20 mol/dm^3. What is K_c for the reaction? _____

To work this problem, you must recall the significance of the coefficients of a balanced equation: they give the mole ratios between reactants and products. Since H_2 and I_2 both have coefficients of 1, it follows that when 0.20 mol/dm^3 of H_2 is produced, an equal number, or 0.20 mol/dm^3, of I_2 must be formed simultaneously. Since 2 mol of HI are required to give 1 mol of H_2, 2(0.20) = 0.40 mol/dm^3 of HI must be consumed to produce 0.20 mol/dm^3 of H_2. Finally, 1.20 - 0.40 = 0.80 mol/dm^3 of HI must be left. Summarizing this reasoning in the form of a table:

	orig. conc. (mol/dm^3)	change (mol/dm^3)	equil. conc. (mol/dm^3)
HI	1.20	-0.40	0.80
H_2	0.00	+0.20	0.20
I_2	0.00	+0.20	0.20

Basic Skills • 205

With this information, K_c is readily calculated.

$$K_c = \frac{[H_2] \cdot [I_2]}{[HI]^2} = \frac{(0.20)(0.20)}{(0.80)^2} = 0.062$$

Problem 15.3 is entirely analogous to the example just worked; 15.13 and 15.29 are similar. Problems 15.14 and 15.30 involve the same principle, but are somewhat more subtle. Note that in these and many other equilibrium problems, you may be given amounts in moles rather than concentrations (mol/dm^3). If that is the case, it may be simplest to set up the equilibrium table in terms of numbers of moles, waiting until the last step to obtain concentrations by dividing through by the volume. However you do it, the distinction between amount and concentration is an important one, which you should have learned in Chapter 12.

5. Given the value of K_c, predict:

a. the direction in which a chemical system will move to reach equilibrium.

The principle involved here is discussed in the text and illustrated in Example 15.3. Notice the importance of distinguishing between "original" or arbitrary concentrations, which may have any value, and equilibrium concentrations, which must be in a fixed ratio given by K_c. The symbol [ ] is used throughout the text for equilibrium concentrations.

This skill is applied in Problems 15.4, 15.15 and 15.31.

b. the equilibrium concentrations of all species, given their initial concentrations.

This is probably the most useful application of the equilibrium constant expression and the one students ordinarily find most difficult. Read carefully Example 15.4 and the comments that follow. It may be worth cautioning against some of the mistakes commonly made in analyzing problems of this type.

- Don't confuse original and equilibrium concentrations. In part (a) of Example 15.4, you might be tempted to set [N$_2$] = [O$_2$] = 0.81 mol/dm^3, which would lead to a quite different (and incorrect!) answer. Remember, since some NO is formed, the concentrations of reactants must decrease; you can't get something for nothing!
- Be sure you relate properly the changes in reactant and product concentrations. Referring again to Example 15.4, students occasion-

206 • 15–Chemical Equilibrium in Gaseous Systems

ally take the change in NO concentration to be x rather than $2x$, forgetting that the coefficients of the balanced equation require a 2:1 mole ratio of NO to N_2.

Calculations of this type are illustrated most simply by Problems 15.5, 15.16, and 15.32. Problems 15.17 and 15.33 are slightly more subtle, but quite simple mathematically if you set them up properly. Problems 15.18a, 15.19, 15.34a, and 15.35 are the most difficult of this type in that they require that you solve a quadratic equation.

6. Using Le Chatelier's Principle, predict the effect of:
   a. adding a reactant or product
   b. changing the volume
   c. changing the temperature
upon the composition of an equilibrium system.

This skill is discussed in Section 15.4 of the text. To summarize, we can say that when:

– a species is added to a system at equilibrium, that reaction occurs which consumes part of that species.
– the volume is increased, the position of the equilibrium shifts so as to increase the number of moles.
– the temperature is increased, the position of the equilibrium shifts so as to favor the endothermic reaction.

The following example illustrates all of these ideas.

Consider the equilibrium: $Cl_2(g) \rightleftharpoons 2 Cl(g)$; $\Delta H = +243$ kJ. Predict what will happen if:
    $Cl(g)$ is added to the equilibrium container. _____
    the volume of the container is increased. _____
    the temperature is increased. _____
Addition of $Cl(g)$ should cause reaction to proceed in the reverse direction (right to left) to consume part of the added Cl, forming more Cl_2. An increase in volume will cause some Cl_2 to dissociate to Cl, thereby increasing the total number of moles. Finally, an increase in temperature will shift the equilibrium to the right, since the forward reaction is endothermic.

Qualitative predictions of this type are called for in Problems 15.6, 15.20, and 15.36. Example 15.5 shows how, by applying Skill 5a, it is possible to make quantitative predictions about the extent to which equilibrium is shifted when a species is added. The same type of calculation is called for in Problems 15.21 and 15.37.

7. **Given K_c at one temperature and ΔH for the reaction, calculate K_c at another temperature, using Equation 15.12.**

This skill is illustrated in Example 15.6. Note the similarity to vapor pressure calculations using the Clausius-Clapeyron equation (Chapter 11). Problems 15.7 and 15.22 are entirely analogous to Example 15.6. In Problem 15.38, the argument is turned around; you are asked to calculate the temperature at which K_c has a certain value.

8. **Relate the equilibrium constant for a reaction to the standard free energy change.**

The basic relationship here is Equation 15.13, applied in Example 15.7 and Problems 15.8, 15.23, and 15.39.

SELF-TEST

True or False

1. A chemical equation describes the relative changes in the ( ) numbers of moles of reactants and products as equilibrium is approached.

2. Equilibrium can be considered as a balance struck between ( ) two opposing tendencies: a system will tend to move toward a state of maximum enthalpy; a system will tend to move toward a state of minimum entropy.

3. The pressure exerted by a sample of gaseous benzene in ( ) equilibrium with liquid benzene will depend on the container volume.

4. Higher temperatures would favor the production of more ( ) product in the following system: benzene (l) $\rightleftharpoons$ benzene (g).

5. The expression for K_c always shows all gaseous or dissolved ( ) species, never pure solid or liquid species.

6. Gaseous hydrogen and oxygen are allowed to react to form ( ) liquid water. The value of K_c for the reaction $2\ H_2(g) + O_2(g) \rightleftharpoons 2\ H_2O(l)$ will depend upon the initial relative amounts of H_2 and O_2.

7. One mole of HI(g) and one mole of H_2(g) are placed in an ( ) evacuated container at 100°C and allowed to come to equilibrium.

As equilibrium is approached, one can be sure that the H_2 concentration will increase. (The reaction: $2\ HI(g) \rightleftharpoons H_2(g) + I_2(g)$.)

8. The value of K_c is expected to increase with temperature ( ) for any reaction that has a negative value of ΔH.

9. Two students study the equilibrium $2\ H_2(g) + O_2(g) \rightleftharpoons$ ( ) $2\ H_2O(l)$; both use the same initial concentrations at the same temperature. One calculates $[H_2]$ using the equation given; the other, using the equation $H_2(g) + \frac{1}{2} O_2(g) \rightleftharpoons H_2O(l)$. Their answers must be the same.

10. For a given reaction, the value of K_c will depend on initial ( ) concentrations.

Multiple Choice

11. The expression for K_c for the equilibrium $C(s) + CO_2(g) \rightleftharpoons$ ( ) $2\ CO(g)$ is

(a) $\dfrac{2\ [CO]}{[C]\ [CO_2]}$ (b) $\dfrac{[CO]^2}{[C]\ [CO_2]}$

(c) $\dfrac{[CO]^2}{[CO_2]}$ (d) $\dfrac{[CO]}{[CO_2]^2}$

12. Approximately stoichiometric amounts of reactants are ( ) mixed in a suitable container. Given sufficient time, the reactants may be converted almost entirely to products if
   (a) K_c is much less than one
   (b) K_c is much larger than one
   (c) $\Delta G = 0$
   (d) ΔG is a large positive number

13. At a certain temperature, $K_c = 1$ for the reaction $2\ HCl(g) \rightleftharpoons$ ( ) $H_2(g) + Cl_2(g)$. In this system at equilibrium, then, one can be sure that
   (a) $[HCl] = [H_2] = [Cl_2] = 1$
   (b) $[H_2] = [Cl_2]$
   (c) $[HCl] = 2 \times [H_2]$
   (d) $\dfrac{[H_2]\ [Cl_2]}{[HCl]^2} = 1$

14. Given the equilibrium constants for the following reactions: ( )

$$2\ Cu(s) + \tfrac{1}{2} O_2(g) \rightleftharpoons Cu_2O(s),\ K_1$$

$$Cu_2O(s) + \tfrac{1}{2} O_2(g) \rightleftharpoons 2\ CuO(s),\ K_2$$

one can show that for

$$2\ Cu(s) + O_2(g) \rightleftharpoons 2\ CuO(s),\ K_c \text{ is equal to}$$

(a) $K_1 + K_2$ (b) $K_2 - K_1$
(c) $K_1 \times K_2$ (d) K_2/K_1

15. What would you predict to be the conditions that would ( ) favor maximum conversion of noxious nitric oxide and carbon monoxide?

$$NO(g) + CO(g) \rightarrow \tfrac{1}{2} N_2(g) + CO_2(g),\ \Delta H = -373\ kJ$$

(a) low T, high P   (b) high T, high P
(c) low T, low P   (d) high T, low P

16. The position of equilibrium would not be appreciably ( ) affected by changes in container volume for
    (a) $H_2(g) + I_2(s) \rightleftharpoons 2\ HI(g)$
    (b) $N_2(g) + O_2(g) \rightleftharpoons 2\ NO(g)$
    (c) $N_2(g) + 3\ H_2(g) \rightleftharpoons 2\ NH_3(g)$
    (d) $H_2O_2(l) \rightleftharpoons H_2O(l) + \tfrac{1}{2} O_2(g)$

17. For the reaction ( )

$$4\ NH_3(g) + 7\ O_2(g) \rightleftharpoons 2\ N_2O_4(g) + 6\ H_2O(l)$$

increasing the pressure by the addition of neon gas would be expected to
    (a) increase the yield of N_2O_4 at equilibrium
    (b) decrease the yield of N_2O_4 at equilibrium
    (c) speed up the reaction in the forward direction
    (d) make no change in the relative amounts of NH_3 and N_2O_4 at equilibrium

18. For this same reaction, a decrease in pressure brought about ( ) by an increase in container volume would be expected to
    (a) increase the yield of water at equilibrium
    (b) diminish the extent of the forward reaction
    (c) increase the value of K_c
    (d) have no effect on the equilibrium attained

19. Which of the following changes will invariably increase the ( ) yield of products at equilibrium?
    (a) an increase in temperature
    (b) an increase in pressure
    (c) addition of a catalyst
    (d) increasing original reactant concentrations

210 • 15–Chemical Equilibrium in Gaseous Systems

20. In which of the following cases will the least time be ( )
required to arrive at equilibrium?
    (a) K_c is very small   (b) K_c is approximately one
    (c) K_c is very large   (d) cannot say

21. In order to reach equilibrium in a shorter time interval, ( ) which one of the following would be appropriate to most any chemical reaction?
    (a) decrease the concentrations of reacting substances
    (b) increase temperature and pressure
    (c) decrease temperature
    (d) use only stoichiometric amounts of reactants

22. For the following reaction, one would predict the form of ( ) K_c to be

$$AgCl(s) + 2\ NH_3(aq) \rightleftharpoons Ag(NH_3)_2^+(aq) + Cl^-(aq)$$

(a) $\dfrac{[Ag(NH_3)_2^+]\ [Cl^-]}{[AgCl]\ [NH_3]^2}$ (b) $\dfrac{[Ag(NH_3)_2^+]\ [Cl^-]}{[NH_3]^2}$

(c) $[Ag^+]\ [Cl^-]$ (d) $\dfrac{1}{[AgCl]}$

23. Into a 1.00 dm³ flask at 400°C are placed one mole of N_2, ( ) three moles of H_2, and two moles of NH_3. If K_c for the following reaction is about 0.5 at 400°C, what reaction, if any, can be expected to occur?

$$N_2(g) + 3\ H_2(g) \rightleftharpoons 2\ NH_3(g)$$

    (a) left to right   (b) right to left
    (c) system is at equilibrium   (d) cannot say

24. For the reaction in (23), the value of K_c is about 0.08 at ( ) 500°C. One can therefore say that
    (a) the reaction is endothermic
    (b) the reaction is exothermic
    (c) K_c is independent of temperature
    (d) K_c is inversely proportional to the absolute temperature

25. If we start with one mole of N_2, three moles of H_2, and two ( ) moles of NH_3 in a 1.00 dm³ container at 500°C (where K_c = 0.08 for the reaction above), at equilibrium
    (a) the number of moles of N_2, H_2, and NH_3 will be in the ratio 1:3:2
    (b) the number of moles of N_2 and H_2 will be in the ratio 1:3

(c) the number of moles of N_2 will be one
(d) the total number of moles will be six

26. How do the equilibrium constants for the forward and ( ) reverse reactions compare to each other?
    (a) they are always the same
    (b) their sum must equal one
    (c) their product is one
    (d) they are not related

27. If a system contains SO_2, O_2, and SO_3 gases at equilibrium, ( )

$$SO_2(g) + \tfrac{1}{2} O_2(g) \rightleftharpoons SO_3(g)$$

an increase in the partial pressure of SO_3 brought about by the addition of more SO_3 to the system will result in
    (a) a reaction in which some SO_3 is formed
    (b) a reaction in which all of the added SO_3 is consumed
    (c) a reaction in which some of the added SO_3 is consumed
    (d) no reaction

28. It is often the case that K_c is very small for a reaction under ( ) almost all conditions. Yet the reaction may be used to produce significant amounts of products. How can this be?
    (a) an alchemist is employed to increase the yield
    (b) the reaction is carried out at very high temperatures
    (c) an alternate series of reactions, giving the same net result, is used
    (d) product is removed from the system as it is formed

29. For the reaction $CaO(s) + CO_2(g) \rightleftharpoons CaCO_3(s)$, we find that ( ) $K_c = 277$ at 800°C. What is the equilibrium concentration of CO_2?
    (a) 277 M
    (b) (1/277) M
    (c) $(277)^{1/2}$ M
    (d) it depends on the amounts of reactants

30. Under certain conditions, the same reaction (29) gives one ( ) mole of CO_2 in the reaction vessel with both solids present at equilibrium. The heat of reaction, for the reaction as written, is −178 kJ. Adding more $CaCO_3(s)$ will have what effect?
    (a) the concentration of CO_2 increases
    (b) the concentration of CO_2 decreases
    (c) no effect, provided $\Delta V_{gas} = 0$
    (d) the effect depends on the temperature

212 • 15–Chemical Equilibrium in Gaseous Systems

31. At a certain temperature, $K_c = 1.2 \times 10^3$ for the (   ) equilibrium $CO(g) + Cl_2(g) \rightleftharpoons COCl_2(g)$. If excess O_2 were added so that any $CO(g)$ present were removed (by conversion to CO_2 and not to $COCl_2$), what would happen to the $COCl_2$ initially present in the equilibrium mixture?
    (a) it would vanish
    (b) more would be formed
    (c) nothing
    (d) its amount would be slightly diminished

32. For the reaction $H_2(g) + Br_2(g) \rightleftharpoons 2 HBr(g)$, $K_c = 4.0 \times 10^{-2}$. For the reaction $HBr(g) \rightleftharpoons \frac{1}{2} H_2(g) + \frac{1}{2} Br_2(g)$, K_c is (   )
    (a) 4.0×10^{-2} (b) 2.0×10^{-1}
    (c) 5.0   (d) 25

33. For the reaction $2 SO_3(g) \rightleftharpoons 2 SO_2(g) + O_2(g)$, $K_c = 32$. If (   ) $[SO_3] = [O_2] = 2.0$ M, then $[SO_2]$ is
    (a) 0.031   (b) 0.25
    (c) 5.7   (d) 8.0

Problems

34. For the reaction $2 HI(g) \rightleftharpoons H_2(g) + I_2(g)$, $K_c = 0.010$ at 500°C. Calculate the concentrations of all species at equilibrium, starting with 0.30 mol of HI in a 5.0 dm³ container.

35. At room temperature, the following equilibrium is established:

$$H_2(g) + S(s) \rightleftharpoons H_2S(g)$$

There are 1.00 mol of H_2 and 1.00 mol of H_2S present at equilibrium with solid sulfur in a 1.00 dm³ flask.
    (a) What is K_c at room temperature?
    (b) The temperature is raised and a new equilibrium is established. An additional 0.67 mol of solid sulfur is formed. What is K_c at the higher temperature?

36. At 1215K, $K_c = 1.44$ for the reaction $CO_2(g) + H_2(g) \rightleftharpoons CO(g) + H_2O(g)$. If one places 0.150 mol of each gas in a 30.0 dm³ flask at 1215K
    (a) in which direction does the reaction proceed? Show by calculation!
    (b) what will be the equilibrium concentrations of CO and H_2O?

*37. Consider the reaction $2\ HI(g) \rightleftharpoons H_2(g) + I_2(g)$.
   (a) At 1000°C, 25% of a sample of HI is found to decompose to the elements at equilibrium. Calculate K_c.
   (b) At this same temperature, only 0.4% of any sample of HBr is decomposed to the elements at equilibrium. Account for the difference in HI and HBr behavior.

*38. For the reaction $C(s) + CO_2(g) \rightleftharpoons 2\ CO(g)$, $K_c = 0.054$ at 1040K and 0.0065 at 940K. From this data alone, calculate:
   (a) K_p at 1040K; at 940K
   (b) ΔH for the reaction
   (c) $\Delta G^{1\ atm}$ for the reaction at 940K
   (d) $\Delta S^{1\ atm}$ for the reaction
Compare ΔH and ΔS to values you would calculate from ΔH_f and $\Delta G_f^{1\ atm}$ at 298K.

SELF-TEST ANSWERS

1. T (An equation does not, for instance, tell us how fast equilibrium is approached or what the equilibrium concentrations will be.)
2. F (Minimum enthalpy, maximum entropy both favor spontaneity.)
3. F (Equilibrium vapor pressure depends only on T — Chapter 11.)
4. T (This kind of prediction was made and explained in Chapter 11.)
5. T (For dissolved species, see Chapter 18.)
6. F (K_c for a given reaction depends only on T.)
7. T (I_2 must form. Why?)
8. F (Decrease. Predict this from Le Chatelier's Principle or Equation 15.12.)
9. T (The molecules don't care what coefficients are used!)
10. F (K_c for a given equation depends only on T.)
11. c (Do not include the solid.)
12. b
13. d (The composition could be anything, provided this ratio equals K_c.)
14. c (Write out all three expressions for K_c; this illustrates the Rule of Multiple Equilibria.)
15. a (Low T favors the exothermic reaction; high P favors fewer moles of gas.)
16. b (No change in the number of moles of gas.)
17. d (Neon takes no part in the reaction.)
18. b (Favoring the larger number of moles of gas.)
19. d (Why?)

214 • 15–Chemical Equilibrium in Gaseous Systems

20. d (See Chapter 16; K_c, like ΔG, says nothing about reaction speed.)
21. b (Again, Chapter 16. Can you explain why high T, P usually increase reaction speed?)
22. b (Dissolved species only! See Chapter 21.)
23. a (Concentration quotient is less than K_c. What would K_c have to be for this system to be already at equilibrium?)
24. b
25. b
26. c (Choose any equation and show this must be so.)
27. c (Apply Le Chatelier's Principle; the number of moles of SO_3 has been temporarily increased. Note that P has not increased for each component.)
28. d (Removal of product shifts equilibrium to form yet more product.)
29. b ($K_c = 1/[CO_2]$.)
30. c
31. a (By decomposing to $CO + Cl_2$. The effect is described in the answer to 28.)
32. c
33. d ($K_c = 32 = [SO_2]^2[O_2]/[SO_3]^2 = [SO_2]^2/2.0$)
34. Let $x = [H_2] = [I_2]$; $[HI] = 0.060 - 2x$

$$\frac{x^2}{(0.060 - 2x)^2} = 0.010; x = 0.0050 = [H_2] = [I_2]; [HI] = 0.050$$

35. (a) $K_c = 1.00$
    (b) $[H_2] = 1.00 + 0.67 = 1.67$
        $[H_2S] = 1.00 - 0.67 = 0.33$ $\Big\}$ $K_c = 0.33/1.67 = 0.20$

36. (a) concentration quotient = 1; forward reaction occurs
    (b)

	original	final
conc. CO	5.00×10^{-3}	$5.00 \times 10^{-3} + x$
conc. H_2O	5.00×10^{-3}	$5.00 \times 10^{-3} + x$
conc. CO_2	5.00×10^{-3}	$5.00 \times 10^{-3} - x$
conc. H_2	5.00×10^{-3}	$5.00 \times 10^{-3} - x$

$$\frac{(5.00 \times 10^{-3} + x)^2}{(5.00 \times 10^{-3} - x)^2} = 1.44; \frac{5.00 \times 10^{-3} + x}{5.00 \times 10^{-3} - x} = 1.20$$

$$x = \frac{1.00 \times 10^{-3}}{2.20} = 0.45 \times 10^{-3}$$

$[CO] = [H_2O] = 5.45 \times 10^{-3}$; $[CO_2] = [H_2] = 4.55 \times 10^{-3}$

*37. (a) 0.028
     (b) HBr is more stable (ΔG_f is more negative at 1000K), primarily reflecting stronger bond in HBr than HI, compared to the elements.

*38. (a) $\Delta n_{gas} = 1$, $K_p = K_c (0.0821\ T)$
         at 1040K, $K_p = 4.6$; at 940K, $K_p = 0.50$

(b) $\log(4.6/0.50) = \dfrac{\Delta H}{2.30\,(8.31)} \times \dfrac{100}{1040 \times 940}$;
$\Delta H = 180 \times 10^3$ J $= 180$ kJ

(c) $\Delta G^{1\ \text{atm}} = -2.30(8.31)T \log K_p = -2.30(8.31)(940)\log 0.50$
$= 5400$ J $= 5.4$ kJ

(d) From Chapter 14:

$$\Delta S^{1\ \text{atm}} = \dfrac{\Delta H - \Delta G^{1\ \text{atm}} \text{ at } 940\text{K}}{940\text{K}} = 0.186 \text{ kJ/K}$$

From Tables 4.1 and 14.2 one calculates: $\Delta H = 172.5$ kJ
$\Delta G^{1\ \text{atm}}_{298\text{K}} = 119.8$ kJ and $\Delta S^{1\ \text{atm}} = 0.177$ kJ/K.

Note: From experimental measurements of the equilibrium constant at two or more temperatures, one can determine ΔH, ΔS, and ΔG for a reaction!

SELECTED READINGS

Alternative discussions of equilibrium and its connections to thermodynamics are given in most of the references of Chapter 14.

Further practice in solving problems can be found in the problem manuals listed in the Preface.

Connecting equilibrium and rates of reaction (Chapter 16) is considered in:

Guggenheim, E. A., More about the Laws of Reaction Rates and Equilibrium, *Journal of Chemical Education* (November 1956), pp. 544–545.

Mysels, K. J., The Laws of Reaction Rates and Equilibrium, *Journal of Chemical Education* (April 1956), pp. 178–179.

16
RATES OF REACTION

QUESTIONS TO GUIDE YOUR STUDY

1. Knowing what reactions *can* occur (criteria have been established in Chapters 14 and 15), can you now say what reactions *will* occur? In particular, can you say when a given reaction will begin, how fast it will go, and when it will stop?

2. What properties of a system could you observe to see how fast a reaction is occurring? What kinds of measurements would you make? How can you measure the rates for very fast and very slow reactions?

3. How does the rate of a reaction depend on conditions such as temperature, pressure, and the physical state of the reactants?

4. How does the rate of a reaction depend on concentration? How do you experimentally arrive at such a relationship (the *rate law*)?

5. How do you account for a particular rate law in terms of molecular rearrangements? (The sum of these *elementary reactions* constitutes the *reaction mechanism*.) Is there more than one possible mechanism?

6. How do you account for the fact that most reactions speed up as temperature is increased? Can you quantitatively relate rate and temperature? (Recall that kinetic theory relates molecular speeds and temperature.)

7. How does the rate of a reaction change with time? (Recall that a reaction is expected to approach equilibrium.)

8. How can you change or control the rate of a reaction? (Any common examples you can think of?) How are reaction rates controlled or altered in living organisms?

9. Can you predict or calculate the rate, or write the rate law, for a particular reaction without ever carrying out that reaction? (Recall that you have been able to calculate ΔH, ΔG, and K_c for such a case.)

218 • 16–Rates of Reaction

10. In what ways are rate laws useful? What, if anything, do they let you say about the feasibility of carrying out a particular reaction?

11.

12.

YOU WILL NEED TO KNOW

Concepts

1. How to interpret the sign of ΔH — Chapter 4.
2. How to interpret the Maxwell distribution of molecular energies and its dependence on temperature — Chapter 5.

Math

1. How to calculate logs and antilogs — Appendix 4.
2. How to work problems involving concentration units — Chapter 12.
3. How to work with the graph of an equation for a straight line (e.g., $\log k = -B/T + A$ has the form $y = ax + b$) — see a math text or Preface.

CHAPTER SUMMARY

From the free energy change (Chapter 14) or the equilibrium constant (Chapter 15), we can determine whether or not a reaction will take place and, if so, the extent to which it will occur, *given sufficient time*. Thermodynamic quantities such as ΔG and K_c are derived from experimental measurements; they require no knowledge of molecular behavior. In contrast, when we deal with the rate of a reaction we try to establish the path or mechanism by which it occurs. This requires that we make some assumptions as to what the molecules are up to. Hopefully, these assumptions can be checked by experimental measurements which tell us how reactant concentrations change with time.

The simplest molecular model for reaction path assumes that, in order for reaction to occur, two high-energy molecules must collide. This implies that reaction rate should increase with concentration; the more molecules there are in a given volume, the more collisions there will be in unit time. Experimentally, we find that rate does indeed increase with concentration.

However, the relationship is more complicated than a simple collision mechanism would imply. If the path of every reaction involved nothing more than a simple, two-molecule collision, all reactions would be second order. That is, in the general rate expression

$$A + B \rightarrow \text{products; rate} = k(\text{conc. A})^m (\text{conc. B})^n$$

we would expect that m + n = 2. Experimentally, we find that reactions of other orders are common. In particular, we frequently encounter first order reactions, for which the rate law takes the form

$$\text{rate} = kX$$

and the relationship between concentration and time is given by the equation

$$\log_{10} \frac{X_0}{X} = \frac{kt}{2.30}$$

where X_0 and X are reactant concentrations at times t = 0 and t, respectively. Reactions of other integral orders (0, or 3) or fractional orders (1/2, 3/2) are also known.

The basic weakness of the simple collision model is that very few reactions appear to proceed by a single step. More frequently, the reaction mechanism involves a series of steps, each of which may be a bimolecular collision or the decomposition of an unstable, high-energy species. In most cases, one step is considerably slower than the others and hence determines both the overall rate and the observed reaction order. Experimental data on reaction rate may suggest what this step is but it can never (?) establish an unambiguous mechanism for a particular reaction.

The collision theory also explains, quite satisfactorily, the dependence of reaction rate upon temperature. If we assume that only collisions between high-energy molecules are effective, we can derive the Arrhenius relation between rate constant and temperature:

$$\log_{10} k = \text{constant} - \frac{E_a}{(2.30) RT}$$

The activation energy, E_a, represents the minimum energy that must be available if a collision is to be fruitful. We see from the Arrhenius equation that a reaction which requires a high activation energy will ordinarily be slow (small rate constant). Conversely, if E_a is very small, most of the molecules have sufficient energy to react when they collide and reaction should occur very rapidly.

220 • 16–Rates of Reaction

It may be possible to lower the activation energy by finding a different reaction path with a lower energy barrier. In the laboratory, we do this by finding an appropriate catalyst for the reaction. Frequently, a catalyst provides an active surface upon which reaction can occur. Remember that a catalyst changes only the activation energy and hence the rate of reaction; it does not affect the overall energy difference between products and reactants and hence cannot change the equilibrium constant (Equation 15.12 relates the overall energy change, ΔH, to the equilibrium constant K).

BASIC SKILLS

1. Determine the order of a reaction given:
   a. the initial rate as a function of concentration of reactants.

This skill is illustrated in Example 16.1. Note (in parts b and c) that data of this type can be used to obtain the rate constant for a reaction and to predict rates at any given concentration.

Problems 16.8 and 16.23 are analogous to Example 16.1. Problems 16.9 and 16.24 involve a similar approach, but the reasoning is turned around; assuming a particular order, you are asked to calculate the rate at a given concentration (16.9) or a rate constant (16.24).

   b. concentration of reactant as a function of time.

See Example 16.3 and Figure 16.4. Essentially what is done here is to determine whether the data fits Equation 16.4 (zero order), 16.5 (first order), or 16.7 (second order). It should be clear from the form of these equations that:
   zero order: plot of conc. vs. t is a straight line.
   first order: plot of log conc. vs. t is a straight line.
   second order: plot of 1/conc. vs. t is a straight line.

2. Use the rate equation for a first order reaction (Equation 16.5) to obtain:
   a. the concentration of reactant after a given time, knowing its original concentration and the rate constant;
   b. the time required for the concentration of reactant to drop to a particular value, given the rate constant and the original concentration.

The use of the first order rate equation for these purposes is illustrated in Example 16.2 parts (a) and (b). Notice that there is some advantage in retaining the quantity "log X_0/X" as a unit in your calculations. Since X_0 is always greater than X, log X_0/X is always a positive quantity. Most students prefer to avoid using negative logarithms whenever possible.

Problems 16.3 and 16.12a are straightforward examples of calculations involving concentration-time relations in first order reactions. Problems 16.13 and 16.28 are slightly more difficult but really involve little more than substituting into the first order rate equation.

3. **Given either the half-life or the rate constant for a first order reaction, calculate the other quantity.**

For a first order reaction, the relation between these two quantities is: $t_{1/2} = 0.693/k$. This relation is derived and then applied in Example 16.2, part (c). See also Problems 16.12b and 16.27.

4. **Use the Arrhenius equation (Equation 16.10) to obtain**

 a. the rate constant at T_2, given its value at T_1 and the activation energy;

 b. the activation energy, given rate constants at two different temperatures;

 c. the temperature at which k will have a specified value, given E_a and k_1 at T_1.

Skills 4a and 4b are illustrated in Example 16.4, parts (a) and (b). The example below applies the same reasoning for 4c.

The activation energy for a certain reaction is 6400 J. The rate constant is $1.00 \times 10^{-2}/s$ at 27°C. At what temperature will k be $2.00 \times 10^{-2}/s$?

Looking at Equation 16.10, we see that we are given all the quantities except T_2.

$$k_1 = 1.00 \times 10^{-2}/s; \quad T_1 = (27 + 273)K = 300K$$

$$k_2 = 2.00 \times 10^{-2}/s; \quad E_a = 6400 \text{ J}$$

Substituting:

$$\log_{10} \frac{2.00 \times 10^{-2}}{1.00 \times 10^{-2}} = \log_{10} 2.00 = \frac{6400 (T_2 - 300)}{(2.30)(8.31)(300)T_2}$$

Noting that $\log_{10} 2.00 = 0.301$ and that $6400/(2.30)(8.31)(300) = 1.12$, we have:

$$0.301 = \frac{1.12 (T_2 - 300)}{T_2}$$

Solving: 0.301 T$_2$ = 1.12 T$_2$ - 1.12 (300); T$_2$ = 410K

Note the similarity between Equation 16.10, which describes how the rate constant varies with temperature, and Equation 15.12 (Chapter 15) which tells us how the equilibrium constant changes with temperature.

Of the problems at the end of Chapter 16 that deal with activation energies, rate constants, and temperatures, Problems 16.4, 16.16, and 16.31 are the simplest. Problems 16.14 and 16.29 are slightly more difficult but much more interesting.

5. Given two of the three quantities E$_a$ (activation energy), E$_a$' (activation energy for reverse reaction), ΔH (enthalpy change), calculate the other quantity.

The appropriate equation here is 16.11 (see also Fig. 16.6). The following example illustrates its use.

For a certain reaction, ΔH = -20 kJ and E$_a$ = 25 kJ. What is the value of E$_a$', the activation energy for the reverse reaction? _____

Solving Equation 16.11 for E$_a$':

E$_a$' = E$_a$ - ΔH = 25 kJ - (-20 kJ) = 45 kJ

See Problems 16.17 and 16.32.

6. Determine whether a proposed mechanism for a reaction is consistent with the observed rate expression.

The general principle here is that the form of the rate expression is determined by the slowest step of the mechanism. This step often involves a very reactive intermediate which is not one of the original reactants. Such is the case with the I atom in Example 16.5. The rate expression observed in the laboratory cannot involve such intermediates; in the example quoted, the rate is expressed in terms of the concentrations of the two main reactants, H_2 and I_2.

This skill is applied in Problems 16.19, 16.20, 16.34, and 16.35. Note that the equations written in these problems represent actual mechanisms. Thus, in Problem 16.20(a), the assumption is that the reaction occurs through a collision of a CO molecule with an NO_2 molecule.

SELF-TEST

True or False

1. The rate of reaction ordinarily decreases with time. ( )

2. If the rate of reaction doubles when the concentration is doubled, the reaction must be first order. ( )

3. For a zero order reaction the rate constant k would have the units s^{-1}. ( )

4. In general, very fast reactions have small activation energies. ( )

5. The activation energy for a reaction can be obtained by taking the difference in enthalpy between reactants and products. ( )

6. An enzyme-catalyzed reaction may be inhibited by adding a substance with a structure very similar to that of the substrate. ( )

7. Any reaction for which the rate-determining step involves a collision between two molecules must be second order. ( )

8. The order of reactions occurring at solid surfaces frequently changes with concentration. ( )

9. It is possible for the activation energy, as calculated from the Arrhenius equation, to be negative. ( )

10. The reaction $H_2(g) + I_2(g) \rightarrow 2\ HI(g)$ is first order in both H_2 and I_2. Consequently, the mechanism must involve a simple bimolecular collision between H_2 and I_2 molecules. ( )

Multiple Choice

11. The rate of a chemical reaction usually varies with ( )
    (a) concentration (b) temperature
    (c) time   (d) all of these

12. For the reaction $2\ NO(g) + O_2(g) \rightarrow 2\ NO_2(g)$, the rate is expressed as $-\Delta$ conc. $O_2/\Delta t$. An equivalent expression would be ( )
    (a) Δ conc. $NO_2/\Delta t$ (b) $-\Delta$ conc. $NO_2/\Delta t$
    (c) -2Δ conc. $NO_2/\Delta t$ (d) none of these

13. For a certain decomposition the rate is 0.30 mol/(dm$^3 \cdot$s) when the concentration of reactant is 0.20 M. If the reaction is second order, the rate [mol/(dm$^3 \cdot$s)] when the concentration is 0.60 M will be ( )
    (a) 0.30   (b) 0.60
    (c) 0.90   (d) 2.7

14. In a certain first order reaction the half life is 20 min. ( )
The rate constant k in min^{-1} is about
   (a) 0.035
   (b) 0.35
   (c) 13.9
   (d) cannot tell

15. For a given reaction of first order, it takes 20 min for ( ) the concentration to drop from 1.0 M to 0.60 M. The time required for the concentration to drop from 0.60 M to 0.36 M will be
   (a) more than 20 min
   (b) 20 min
   (c) less than 20 min
   (d) cannot tell

16. First order rate constants are 0.024 a^{-1} for radioactive ( ) ^{90}Sr and 0.13 a^{-1} for ^{60}Co. How do the half-lives compare?
   (a) t½ of Sr is the longer
   (b) t½ of Sr is the shorter
   (c) they are the same
   (d) they cannot be compared

17. For a first order reaction, a straight line is obtained if you ( ) plot
   (a) log conc. vs. time
   (b) conc. vs. time
   (c) 1/conc. vs. time
   (d) log conc. vs. 1/time

18. The activation energy of a certain reaction is 15 kJ. The ( ) activation energy for the reverse reaction is
   (a) −15 kJ
   (b) > 15 kJ
   (c) < 15 kJ
   (d) cannot tell

19. The effectiveness of a catalyst depends upon its ability to ( )
   (a) decrease the activation energy
   (b) increase K$_c$
   (c) increase reactant concentration
   (d) increase temperature

20. The principal reason for an increase in reaction rate with ( ) an increase in temperature is
   (a) molecules collide more frequently at high temperatures
   (b) the pressure exerted by reactant molecules increases with T
   (c) the activation energy decreases with an increase in T
   (d) the fraction of high-energy molecules increases with T

21. Which of the following will generally speed up a reaction? ( )
   (a) raising the temperature
   (b) increasing reactant concentration
   (c) stirring
   (d) all of these

22. For the chain reaction between H_2 and F_2 to form HF, the ( )
step $H + F \to HF$ represents
   (a) chain initiation
   (b) chain propagation
   (c) chain termination
   (d) the overall mechanism of the reaction

23. The decomposition of ozone is believed to occur by the ( )
mechanism

$$O_3 \rightleftarrows O_2 + O \quad \text{(fast)}$$

$$O + O_3 \to 2 O_2 \quad \text{(slow)}$$

When the concentration of O_2 is increased, the rate will
   (a) increase   (b) decrease
   (c) stay the same   (d) cannot say

24. Enzyme-catalyzed reactions resemble surface reactions most ( )
closely in
   (a) mechanism   (b) E_a
   (c) ΔG (d) ΔH

25. In the two-step reaction $A \to B \to C$, the concentration of ( )
the intermediate B is likely to
   (a) increase steadily with time
   (b) decrease steadily with time
   (c) be independent of time throughout the reaction
   (d) remain constant through most of the reaction

26. Which of the following statements is true for all zero order ( )
reactions?
   (a) the activation energy is very low
   (b) the concentration of reactant does not change with time
   (c) the rate constant, k, is zero
   (d) the rate is independent of time

27. The following mechanism is proposed for the oxidation of ( )
iodide ion, I^-, to iodine:

$$NO + \frac{1}{2} O_2 \to NO_2$$

$$NO_2 + 2 I^- + 2 H^+ \to NO + I_2 + H_2O$$

$$I_2 + I^- \to I_3^-$$

A catalyst in this reaction is
- (a) NO
- (b) I⁻
- (c) O_2
- (d) H⁺

28. To determine the order with respect to I⁻ in the reaction (   ) $2\ H^+(aq) + 3\ I^-(aq) + H_2O_2(aq) \rightarrow 2\ H_2O + I_3^-(aq)$, one would prepare several solutions differing in
- (a) conc. I⁻
- (b) conc. H⁺, conc. H_2O_2
- (c) conc. I⁻, conc. H⁺, conc. H_2O_2
- (d) none of the above; the order is third, as given by the coefficient in the balanced equation.

Problems

29. At 27°C, the first order rate constant for the decomposition of N_2O_5 is $4.0 \times 10^{-4}\ s^{-1}$.
    - (a) What is the rate of decomposition of N_2O_5 when its concentration is $1.8 \times 10^{-3}\ mol/dm^3$?
    - (b) How much N_2O_5 will have reacted in 350 s, starting with $2.0 \times 10^{-3}\ mol/dm^3$?

30. The gas phase reaction $SO_2Cl_2(g) \rightarrow SO_2(g) + Cl_2(g)$ is found to be first order with a rate constant at 320°C of $2.2 \times 10^{-5}\ s^{-1}$.
    - (a) How long will it take for half of a 10.0-g sample of SO_2Cl_2 to decompose at this temperature?
    - (b) Calculate the fraction of a 10.0-g sample of SO_2Cl_2 that will remain unreacted after two hours at 320°C.

31. For the reaction $X \rightarrow 2\ Y$, ΔH is 16.0 kJ; the activation energy for the reverse reaction is 5.5 kJ. Calculate
    - (a) the activation energy for the forward reaction.
    - (b) the rate constant at 45°C given that k at 0°C = $1.06 \times 10^{-5}\ min^{-1}$.

*32. (Calculus problem) Show that for a first order reaction, the amount of reactant X at time t is related to the initial amount X_0 at time zero by the equation $X = X_0 e^{-kt}$. What would a plot of X versus t look like? Interpret X_0 and k in terms of the graph. How would you calculate an instantaneous rate, given k, X_0, and the time?

*33. A certain author writes about 1800 words a day under normal conditions (37°C). His editor makes a few phone calls, applying a little heat, and the author is now seen feverishly (39°C) at work, cranking out 3500 words a day. Estimate the energy barrier separating the undisciplined author from his minor masterpiece.

SELF-TEST ANSWERS

1. T (Rate decreases as reactants disappear.)
2. T (Rate = k(conc.), with a direct proportion.)
3. F (mol/(dm^3·s).)
4. T (So that the majority of reactant molecules have sufficient energy for effective collisions.)
5. F (ΔH is the difference in activation energies for forward and reverse reactions.)
6. T (So as to occupy the active sites on the enzyme molecule.)
7. F (Only if the molecules that collide are two reactants.)
8. T (E.g., from first, at low conc., to zero, at high.)
9. T (Some reactions slow down at T rises.)
10. F (Cannot deduce mechanism unambiguously from rate law.)
11. d
12. d (1/2Δconc. NO$_2$/Δt, or $-$1/2Δconc. NO/Δt.)
13. d (For conc. increase of X3, rate increases X(3)^2 = 9.)
14. a (k = 0.693/t$_½$.)
15. b (Δt is constant for conc. to change by same fractional amount.)
16. a (t$_½$ = 0.693/k; for small k, t$_½$ is large.)
17. a
18. d (Need to know ΔH.)
19. a
20. d (The increase is exponential. See Chapter 5 for effect of ΔT on distribution of energies.)
21. d (Why stirring?)
22. c (Nothing further happens to HF, the product.)
23. b (Conc. O inversely related to that of O$_2$ by the rapidly established equilibrium.)
24. a
25. d (Has to be zero at beginning and end.)
26. d (Except when reactants are finally gone.)
27. a (No net change in the NO.)
28. a (A general principle in any experiment: to determine the effect of factor X on a system, change only X. Any change in the system is most likely attributable to ΔX.)
29. (a) 4.0×10^{-4} s^{-1} $\times$ 1.8×10^{-3} $\dfrac{\text{mol}}{\text{dm}^3}$ = 7.2×10^{-7} $\dfrac{\text{mol}}{\text{dm}^3 \cdot \text{s}}$

    (b) $\log \dfrac{2.0 \times 10^{-3}}{X} = \dfrac{(4.0 \times 10^{-4})(350)}{2.30} = 0.061$

    $\dfrac{2.0 \times 10^{-3}}{X} = 1.15$; X = 1.7×10^{-3} M; 0.3×10^{-3} reacted

30. (a) $t_{1/2} = \dfrac{0.693 \text{ s}}{2.2 \times 10^{-5}} = 3.2 \times 10^4 \text{ s}$

(b) $\log \dfrac{10.0}{X} = \dfrac{(2.2 \times 10^{-5})(7200)}{2.30} = 0.069; \dfrac{10.0}{X} = 1.17;$

$X = 8.55$ g; 85.5% unreacted

31. (a) $16.0 \text{ kJ} + 5.5 \text{ kJ} = +21.5 \text{ kJ}$

(b) $\log \dfrac{k}{1.06 \times 10^{-5}} = \dfrac{21\,500(45)}{(2.30)(8.31)(273)(318)} = 0.583$

$k = (1.06 \times 10^{-5}/\text{min})(3.83) = 4.06 \times 10^{-5} \text{ min}^{-1}$

*32. X_0 is an intercept; k is related to the steepness of the exponential curve.

Instantaneous rate $= -dx/dt = kX_0 e^{-kt}$

*33. $\log \dfrac{3500}{1800} = \dfrac{E_a}{2.30\,(8.31)} \left(\dfrac{2}{312 \times 310} \right)$

$E_a = 270$ kJ (Rather large — it's amazing that anything gets done!) [If this is not so amusing, consider the problem we could have written on E_a: Estimate the activation energy for cricket song if the rate of chirping is 20 min^{-1} at 10°C and 120 min^{-1} at 20°C. Answer: 120 kJ/cricket.]

SELECTED READINGS

An alternative discussion, at about the same level, is given in:

King, E. L., *How Chemical Reactions Occur: An Introduction to Chemical Kinetics and Reaction Mechanisms*, New York, W. A. Benjamin, 1964.

Mechanisms of reactions are discussed in:

Edwards, J. O., From Stoichiometry and Rate Laws to Mechanisms, *Journal of Chemical Education* (June 1968), pp. 381–385.

Jones, M., Jr., Carbenes, *Scientific American* (February 1976), pp. 101–113.

Wolfgang, R., Chemical Accelerators, *Scientific American* (October 1968), pp. 44–52.

Some special effects and problems in kinetics are discussed in the articles by Guggenheim and Mysels, listed in Chapter 15, in the book by Campbell (Chapter 14), and in:

Bunting, R. K., Periodicity in Chemical Systems, *Chemistry* (April 1972), pp. 18–20.

Tamaru, K., New Catalysts for Old Reactions, *American Scientist* (July-August 1972), pp. 474–479.

Winfree, A. T., Rotating Chemical Reactions, *Scientific American* (June 1974), pp. 82–95.

17

THE ATMOSPHERE

QUESTIONS TO GUIDE YOUR STUDY

1. What are some of the bulk properties of the atmosphere? What are its dimensions and mass? Can the Ideal Gas Law be applied to relate some of these properties?

2. What is the overall composition of the atmosphere? How does the composition vary with weather, geographical location, and altitude? What kinds of experiments give this information? What kinds of explanations do we give?

3. How may the components of the atmosphere be separated and identified?

4. What are the origins of the atmospheric components which are most abundant? (For example: What reactions are known to give rise to atmospheric oxygen? To deplete it?)

5. What reactions do the components of the atmosphere take part in, within the atmosphere itself, as well as at the interface of atmosphere and earth and between the atmosphere and living organisms?

6. What are the rates and extents of these reactions?

7. What substances can be considered as atmospheric contaminants? Where do they come from? Where do they go? What effects do they have on atmospheric properties? On life and other processes?

8. How do you test for pollutants? How do you specify their concentrations? How do you control them? How do you prevent them?

9. How do you measure the extent of pollution and its change with time? (Is pollution increasing?)

10. What are some of the reactions by which pollutants can be removed? What are their limitations? Their costs?

11.

12.

YOU WILL NEED TO KNOW

Concepts

    1. How to draw Lewis structures — Chapter 8.
    2. How to interpret the sign and relative magnitudes of thermodynamic quantities (ΔH, ΔS, ΔG) and relate these to changes at the atomic-molecular level — Chapters 4, 14.
    3. How to recognize classes of organic compounds — Chapter 10.

Math

    1. How to write simple balanced equations — Chapter 3.
    2. How to define as well as work with partial pressure; how to work with the Ideal Gas Law — Chapter 5.
    3. How to relate the wavelength and energy of light — Chapter 6.
    4. How to work with concentration units introduced thus far — Chapter 12.
    5. How to predict, qualitatively as well as quantitatively, the effects of changes in reaction conditions (T, P, concentration, nature of reactant, etc.) on:

        spontaneity of reaction (sign of ΔG) — Chapter 14.
        equilibrium concentrations — Chapter 15.
        rate of reaction — Chapter 16.

CHAPTER SUMMARY

    In this chapter, we have attempted to tie together the concepts of chemical kinetics, chemical equilibrium, and chemical bonding introduced in previous chapters, and to discuss the chemical and physical properties of the atmosphere. Interwoven with these concepts is a considerable amount of descriptive chemistry of the major constituents of the atmosphere: nitrogen, oxygen, carbon dioxide, water vapor, and the noble gases.
    Our interest in the chemistry of the very unreactive element nitrogen centers upon the process of nitrogen fixation in which N_2 is converted to

useful compounds. This process is carried out in nature by certain clever bacteria found in the roots of clover and other legumes. Industrially, it is accomplished by reacting nitrogen with hydrogen to form ammonia (Haber process). Ammonia, either directly or in the form of its salts (e.g., NH_4NO_3), is used as a fertilizer. It can be converted by the Ostwald process to nitric acid, which in turn is used to make fertilizers (inorganic NO_3^- salts) and explosives (organic nitro compounds).

Oxygen reacts with all but a very few elements. With most metals, the O_2 molecule is converted to the oxide ion, O^{2-}. Certain of the 1A and 2A metals normally yield peroxides (O_2^{2-}) or superoxides (O_2^-). Perhaps the most important of the reactions of oxygen with the nonmetals is that with sulfur; the sulfur dioxide produced by combustion (primarily of coal and oil) under ordinary conditions is converted first to sulfur trioxide and then to sulfuric acid. When an element forms more than one compound with oxygen, it is ordinarily the "higher" oxide (e.g., CuO, CO_2) which is formed at low temperatures in the presence of excess oxygen ("oxidizing atmosphere"). High temperatures and limited amounts of oxygen, a "reducing atmosphere," favor the lower oxide (e.g., Cu_2O, CO).

In the upper atmosphere, we find many species, such as O_3, O, and O_2^+, which are unstable under ordinary laboratory conditions. These are formed by absorption of high-energy radiation in the ultraviolet and far UV regions of the spectrum; as a result of this high-altitude absorption, very little of this radiation reaches the surface of the earth. Since the reactions that form these species are nonspontaneous from a thermodynamic point of view, we might expect them to be readily reversed. However, because the concentrations of these atoms and ions are extremely low, their rate of recombination to form stable molecules such as O_2 is quite slow.

Among the "minor" but objectionable constitutents of the atmosphere are the oxides of sulfur (mostly SO_2), the oxides of nitrogen (mostly N_2O), carbon monoxide, hydrocarbons, and suspended particles. Even though these species are present at a level of a few parts per million or less, they pose serious problems to human health. The automobile is a major source of three of these five pollutants (NO_X, CO, hydrocarbons) and is largely responsible for smog formation, which involves reaction of NO_2 with hydrocarbons. The combustion of fuels in power plants and industry is responsible for much of the sulfur oxides and suspended solids that enter the atmosphere.

Efforts to reduce the level of air pollution have concentrated largely upon modifications of the internal combustion engine and exhaust system of the automobile. Several different approaches are under study but none, at least at present, seem likely to reduce NO_X and hydrocarbon emissions to the level required to prevent smog formation. Perhaps the ultimate answer will be to use an energy source other than gasoline; possibilities include natural gas, electricity, and steam.

232 • 17–The Atmosphere

BASIC SKILLS

Very few, if any, new concepts are introduced in this chapter. If you examine the problems, you will find that most of them fall into one of three categories.

    a. *The application of concepts introduced in previous chapters*, particularly Chapters 4 and 14 (Thermodynamics), 15 (Chemical Equilibrium), and 16 (Chemical Kinetics). For example, Problems 17.9 and 17.27 are similar to problems in Chapter 14; Problems 17.10 and 17.28 could well have appeared in Chapter 15; Problems 17.14–17.16 and 17.32–17.34 apply the principles of kinetics (Chapter 16).

    b. *Descriptive chemistry, involving methods of preparation or equation writing.* Problems 17.2, 17.6, 17.12, 17.24, and 17.30 are of this type. You should be able to work them with little difficulty if you have assimilated the descriptive material in this chapter, particularly in Sections 17.2 and 17.3. See also Example 17.2.

    c. *Discussion-type questions* based on qualitative ideas introduced in this chapter. See Problems 17.5, 17.11, 17.18, 17.21, and the corresponding answered problems in the right-hand columns.

The two "skills" listed below are quite simple, involving nothing more than conversions (Skill 1) or substitution into an algebraic equation (Skill 2).

1. Using conversion factors given in Table 17.2, convert concentrations of gaseous species from one unit to another.

This skill is illustrated in Example 17.1, which covers virtually every conversion of this type. See also Problems 17.1, 17.4, and 17.22. Incidentally, it is not necessary to memorize the conversion factors given in Table 17.2. All of them follow logically, either from the definitions of the various terms or from the gas laws (Chapter 5). The only one that might puzzle you is the factor relating concentration (mol/dm^3) to partial pressures. Recall that the Ideal Gas Law states that PV = nRT. Hence:

$$\frac{n}{V} = \frac{mol}{dm^3} = \frac{P}{RT} = \frac{P}{(8.31)T} \text{ (P in kilopascals, T in K)}$$

2. Given two of the three quantities relative humidity, partial pressure of water vapor in the air, and equilibrium vapor pressure of water, calculate the third quantity.

Calculations of this type follow directly from the definition of relative humidity (Equation 17.32). See Example 17.3 and Problem 17.3.

SELF-TEST

True or False

1. Helium is the most abundant noble gas in the atmosphere. ( )
2. Of all the gases in the atmosphere, nitrogen is the least reactive. ( )
3. The reaction of oxygen with a 1A metal may produce O^{2-}, O_2^{2-} or O_2^- ions. ( )
4. Carbon dioxide is obtained commercially by fractional distillation of air. ( )
5. When a sample of air is allowed to warm up without changing the total water content, the relative humidity increases. ( )
6. As one moves up in the atmosphere, the ratio (conc. O)/(conc. O_2) increases. ( )
7. A temperature profile of the atmosphere, from sea level up to, say, 200 km could, at least in principle, be obtained by taking readings on a mercury-in-glass thermometer. ( )
8. The ratio of O_3 to O concentration in the upper atmosphere can be calculated knowing the concentration of O_2 and the equilibrium constant for the reaction $O_3 \rightleftarrows O_2 + O$. ( )
9. In order to reduce the concentration of NO in automobile exhaust, it is desirable to increase the temperature at which the fuel is burned. ( )
10. A water solution of MgO is expected to contain an appreciable concentration of $OH^-(aq)$. ( )

Multiple Choice

11. In a synthetic atmosphere made up of He and O_2, the mole fraction of He is 0.80. The mass % of He is ( )
    (a) 80   (b) less than 80
    (c) greater than 80   (d) cannot tell

12. The formula of calcium nitride is ( )
    (a) CaN   (b) Ca_2N
    (c) Ca_2N_3 (d) Ca_3N_2

234 • 17-The Atmosphere

13. In the compound Fe_3O_4, the mole ratio of Fe^{2+} to Fe^{3+} is ( )
    (a) 1:1
    (b) 1:2
    (c) 2:1
    (d) 3:4

14. In the Haber process for making ammonia, high pressures ( ) are used to
    (a) increase the yield, leaving the rate unchanged
    (b) increase the rate, leaving the yield unchanged
    (c) increase the yield and the rate
    (d) increase the equilibrium constant and the rate

15. Consider the reaction $SnO(s) + 1/2\ O_2(g) \rightarrow SnO_2(s)$. This ( ) reaction would be expected to
    (a) become nonspontaneous at high temperatures
    (b) become nonspontaneous at high pressures
    (c) become nonspontaneous at low temperatures
    (d) not occur at any temperature or pressure

16. To increase the yield of SO_3 from SO_2 in the reaction with ( ) O_2, we could
    (a) increase the temperature
    (b) use more SO_2
    (c) add a catalyst
    (d) increase the pressure

17. Of the following noble gases, which reacts most readily with ( ) fluorine?
    (a) He
    (b) Ne
    (c) Kr
    (d) Xe

18. Dry ice is effective in seeding clouds because ( )
    (a) CO_2 and H_2O have similar crystal structures
    (b) it increases the water content of the cloud
    (c) CO_2 molecules offer a nucleus for condensation
    (d) upon sublimation, it lowers the temperature of the water

19. Automobile emissions are not a major source of ( )
    (a) NO
    (b) CO
    (c) hydrocarbons
    (d) SO_2

20. Which one of the following hydrocarbons would be most ( ) likely to contribute directly to smog formation?
    (a) CH_4
    (b) C_2H_4
    (c) C_3H_8
    (d) C_6H_6

21. Of the following fuels, which one, under normal conditions, ( ) produces the lowest concentration of pollutants?
    (a) coal
    (b) wood
    (c) natural gas
    (d) petroleum

22. The superoxide ion, found in KO_2, has the structure ( )
    (a) :Ö—Ö:
    (b) :Ö=Ö:
    (c) :Ö—Ö:
    (d) :Ö—Ö:

23. Of the species $O^{2-}, O_2, O_2^-, O_2^{2-}$,
    (a) which are paramagnetic?_____
    (b) which contain covalent bonds?_____
    (c) which react with water to form H_2O_2?_____
    (d) which react with water to form O_3?_____
    (e) which one is most stable to thermal decomposition?_____

24. Of the species O_2, N_2, O, N, N^+,
    (a) which is most abundant in the atmosphere?_____
    (b) which are unstable under ordinary laboratory conditions?_____
    (c) which is most difficult to form from an energy standpoint?_____
    (d) which are held together by covalent bonds?_____

25. Of the molecules $NO_2, SO_2, O_3, CO, CO_2$,
    (a) which are major contributors to photochemical smog?_____
    (b) which can be converted to strong acids?_____
    (c) which one is most abundant in automobile exhaust?_____
    (d) which are thermodynamically unstable in the presence of O_2?_____

26. Liquefaction of air involves ( )
    (a) an application of the Ideal Gas Law
    (b) the increase of pressure on a sample of air at constant temperature
    (c) departure from ideal behavior
    (d) expansion of air at constant temperature

27. The fact that the metal oxide Mn_2O_7 is a liquid above ( ) $-20°C$ suggests that the bonding is predominantly
    (a) metallic
    (b) ionic
    (c) covalent
    (d) dispersion forces

28. CO emissions may be decreased by ( )
    (a) burning fuels at higher T
    (b) mixing combustion reactants more thoroughly
    (c) passing combustion products over hot charcoal
    (d) all the above

29. Carbon monoxide seems to be removed from the atmos- (   )
phere mostly by
   (a) dissolving in the oceans
   (b) reacting further with O_2 to give CO_2
   (c) thermal decomposition to C and O_2
   (d) consumption by soil microorganisms

Problems

30. The partial pressure of water vapor in the air is 1.60 kPa when the temperature is 24°C and the total pressure is 100 kPa (vp water at 24°C = 2.98 kPa).
   (a) What is the relative humidity?
   (b) What is the mole fraction of water in the air?
   (c) What is the concentration of water in the air (mol/dm^3)?

*31. In actual practice, how does one measure the concentration of nitrogen oxides in automobile exhausts? Are chemical or physical processes involved?

*32. One frequently encounters discussions of the carbon and nitrogen cycles in nature. Consider now the sulfur cycle, at least that part which involves the atmosphere.
   What compounds of sulfur enter the atmosphere? Where do they come from? By what means are they removed from the atmosphere? What are their effects, harmful and beneficial?
   Indicate amounts and concentrations, if known; write equations for reactions whenever possible.

SELF-TEST ANSWERS

1. F (Argon.)
2. F (Noble gases.)
3. T (Resulting from combustion in O_2: Li_2O, Na_2O_2, KO_2.)
4. F (Fractional distillation is used mainly for H_2, O_2, and the noble gases. What would you expect the source to be?)
5. F (More vapor could be present at higher T, since vp increases with T, so RH decreases.)
6. T (Where does the energy required to break the O—O bond come from?)
7. F (The T at high altitudes would have to be calculated from molecular speeds; see Chapter 5. Why?)

Self-Test Answers • 237

8. F (Molecules too far apart, react too slowly to establish equilibrium.)
9. F (With $\Delta H_f > 0$, its formation becomes more extensive at high T. See Chapter 15, relating K_c and T.)
10. T (Basic anhydride; for more on properties of bases, see Chapter 19.)
11. b (He weighs less than O_2.)
12. d (Ions expected are Ca^{2+}; and, if N is to gain electrons, three are required for the octet in N^{3-}. See Chapter 8.)
13. b (1 Fe^{2+}, 2 Fe^{3+} to 4 O^{2-} required for electroneutrality.)
14. c (Higher P favors fewer moles of gas; higher P effectively increases concentration, hence rate. See Chapters 15 and 16, respectively.)
15. a (Consider effect of T for $\Delta S < 0$: $\Delta G = \Delta H - T\Delta S$.)
16. d (Is b as good an answer?)
17. d (Why is Xe most reactive? See the Readings in Chapter 8.)
18. d (Heat is absorbed in this process − Chapter 11.)
19. d (SO_2 comes mainly from burning sulfur-containing coal and oil in power plants.)
20. b (Reactive double bond; draw the Lewis structure − Chapter 8.)
21. c
22. a (One extra valence electron for −1 charge.)
23. (a) O_2, O_2^-, as explained by MO theory, Chapter 8.
    (b) O_2, O_2^-, O_2^{2-}
    (c) O_2^-, O_2^{2-}
    (d) none
    (e) O^{2-} (others have bonds that can be broken)
24. (a) N_2
    (b) O, N, N^+ (Why would you expect them to be unstable at 25°C, 1 atm?)
    (c) N^+ (requiring N≡N bond breaking plus ionization)
    (d) O_2, N_2
25. (a) NO_2, O_3
    (b) NO_2, SO_2 (HNO_3 and H_2SO_4 − see Chapter 19 about acid "strength")
    (c) CO_2 (usual combustion product for carbon compounds − Chapter 10)
    (d) NO_2, SO_2, O_3, CO (giving NO, SO_3, O_2, CO_2)
26. c (Attractive forces *do* exist between gas molecules. Would b or d work, by themselves?)
27. c (Low mp is characteristic of molecular substances − Chapter 9.)
28. b (The others would favor CO formation.)
29. d (See Readings to answer this!)

30. (a) $\dfrac{1.60}{2.98} \times 100 = 53.7\%$

(b) $\dfrac{1.60}{100} = 0.0160$

(c) $\dfrac{n}{V} = \dfrac{P}{RT} = \dfrac{(1.60)}{(8.31)(297)} = 6.48 \times 10^{-4}$ M

*31. You may wish to consult your state or federal environmental protection agency.

*32. A place to start: Kellogg, W. W., The Sulfur Cycle, *Science* (February 11, 1972), pp. 587–595.

SELECTED READINGS

Looking at the atmosphere and its relation to the rest of our planet (or some other planet), setting a perspective:

Herbig, G. H., Interstellar Smog, *American Scientist* (March-April 1974), pp. 200–207.

Huntress, W. T., Jr., The Chemistry of Planetary Atmospheres, *Journal of Chemical Education* (April 1976), pp. 204–208.

Kellogg, W. W., The Sulfur Cycle, *Science* (February 11, 1972), pp. 587–595.

Kenyon, D. H., *Biochemical Predestination*, New York, McGraw-Hill, 1969.

Lewis, J. S., The Atmosphere, Clouds and Surface of Venus, *American Scientist* (September-October 1971), pp. 557–566.

The Biosphere, *Scientific American* (September 1970).

Turner, B. E., Interstellar Molecules, *Scientific American* (March 1973), pp. 51–69.

More specifically oriented toward ecological problems:

Carbon Monoxide: Natural Sources Dwarf Man's Output, *Science* (July 28, 1972), pp. 338–339.

Fennelly, P. F., The Origin and Influence of Airborne Particulates, *American Scientist* (January-February 1976), pp. 46–56.

Stoker, H. S., *Environmental Chemistry: Air and Water Pollution*, Glenview, Ill., Scott, Foresman, 1972.

And a very important process is thermodynamically studied:

Safrany, D. R., Nitrogen Fixation, *Scientific American* (October 1974), pp. 64–80.

18
PRECIPITATION REACTIONS

QUESTIONS TO GUIDE YOUR STUDY

1. What occurs during a precipitation reaction? What would you observe? (Can you give any common example?)

2. Why does precipitation occur? What are the driving forces behind the reaction? What are the reactant particles doing?

3. What substances participate in this class of reaction? Can some generalizations be made? Are there correlations with electronic structure, and hence with the Periodic Table?

4. How do reaction conditions such as temperature and concentration affect the spontaneity and extent of a precipitation reaction?

5. Can you apply Le Chatelier's Principle to determine the effect of changes in reaction conditions? Can such predictions be made quantitative?

6. Can you predict the direction and extent for this kind of reaction? (How have you predicted spontaneity and extent, qualitatively and quantitatively, for other systems?)

7. How do you write and interpret chemical equations for this class of reaction?

8. What can you say about the rates of precipitation reactions? How, for example, would you expect them to depend on temperature or concentration? How do they compare with other reaction rates?

9. What can you say about the macroscopic nature of the products of a precipitation (for example, the shape and size of crystals) and how it depends on reaction conditions?

10. What are some applications for this kind of reaction?

11.

12.

YOU WILL NEED TO KNOW

Concepts

1. How to write and interpret (balanced) chemical equations — Chapters 3 (text and this guide) and 18.
2. The general principles of solubility, and how solubility may change with conditions (e.g., temperature) — Chapters 12, 13.
3. The particle nature of electrolyte solutions (What is an electrolyte? How many ions form per mole of solute?) — Chapter 13.
4. How to define and interpret an equilibrium constant — Chapter 15.

Math

1. How to perform stoichiometric calculations, particularly those involving concentration units — Chapters 3, 12 (and more worked examples in Chapter 18).
2. How to write the equilibrium constant expression for any reaction — Chapter 15.
3. How to predict qualitatively and quantitatively the effects of changes in reactant or product concentrations on the position of equilibrium — Chapter 15.

CHAPTER SUMMARY

This is the first of six chapters devoted to a discussion of the reactions of ions and molecules in water solution. Of the various types of reactions we will consider (precipitation, acid-base, complex ion formation, oxidation-reduction), precipitation is perhaps the simplest. A precipitation reaction will occur when two electrolyte solutions are mixed, if one of the possible products is insoluble. To illustrate: when solutions of $MgCl_2$ and $NaOH$ are mixed, one of the two possible products, $Mg(OH)_2$, is insoluble (what is the other possible product?), so the following reaction occurs:

$$Mg^{2+}(aq) + 2\ OH^-(aq) \rightarrow Mg(OH)_2(s)$$

Notice that the equation written for the reaction includes only those species which actually participate in it; "spectator" ions (Na^+, Cl^-) are omitted.

In order to decide when a precipitation reaction will occur, we need to know water solubilities of various electrolytes. Unfortunately, these cannot be predicted with any confidence from first principles. At the present state of electrolyte solution theory (Chapter 13), the best we can do is to rationalize after the fact; i.e., "explain" why $Mg(OH)_2$ should be much less soluble than NaCl. We have to resort to solubility rules (based simply on observation) such as those listed in Table 18.1 to predict the spontaneity of precipitation reactions.

Table 18.1 lists those compounds which can be expected to precipitate when 0.1 M solutions of the corresponding ions are mixed. If the solutions used are more dilute (and indeed for any calculations) we must know the solubility product of the electrolyte involved. For $Mg(OH)_2$ we find experimentally that

$$K_{sp} = [Mg^{2+}][OH^-]^2 = 1 \times 10^{-11}$$

This is the equilibrium constant expression for the reaction

$$Mg(OH)_2(s) \rightleftarrows Mg^{2+}(aq) + 2\, OH^-(aq)$$

and can be manipulated in every way like the equilibrium constants of Chapter 15. In words, this equation tells us that $Mg(OH)_2$ will precipitate from any solution in which the product of the concentration of Mg^{2+}, times that of OH^- squared, exceeds 1×10^{-11}. Thus, $Mg(OH)_2$ is sufficiently insoluble to precipitate not only when 0.1 M solutions of Mg^{2+} and OH^- are mixed, but also with 0.01 M or even 0.001 M solutions. (Would $Mg(OH)_2$ form if the concentrations of both ions were 0.0001?) A knowledge of the solubility product also enables us to carry out many other practical calculations.

Precipitation reactions are used for a variety of analytical and preparative purposes. For example, we might take advantage of the water insolubility of $Mg(OH)_2$ to:
a. detect Mg^{2+} in a mixture with other cations, none of which forms an insoluble hydroxide;
b. analyze quantitatively for Mg^{2+} by weighing the $Mg(OH)_2$ produced (or the MgO formed by heating it) when an excess of OH^- is added to a solution;
c. separate Mg^{2+} from sea water (this is the ultimate source of the magnesium metal in use today);
d. convert a soluble hydroxide (e.g., CsOH) to another soluble salt of that cation (e.g., CsCl, Cs_2SO_4) by adding the appropriate salt of Mg^{2+} (e.g., $MgCl_2$, $MgSO_4$), filtering off the precipitated $Mg(OH)_2$, and evaporating the remaining solution.

BASIC SKILLS

1. **Write net ionic equations to represent precipitation reactions.**

The reasoning involved here is described in some detail in Example 18.2 and applied in Problems 18.1, 18.7, and 18.26. Note that in order to follow the example or work the problems you must be familiar with the solubility rules given in Table 18.1.

The phrase "net ionic equation" introduced in this chapter is perhaps a misnomer; "net equation" might be better. Such equations can involve the formulas of insoluble solids (e.g., $PbCl_2$) and, as we shall see in later chapters, molecules in water solution (e.g., HF, H_2O). The essential point is that in any equation, *only those species which actually take part in the reaction are included.*

2. **Given the equation for a precipitation reaction, relate the amounts (numbers of moles or grams) of reactants and products.**

This skill is essentially identical with that first discussed in Chapter 3 in connection with mass relations in chemical reactions. Here for the first time it is applied to reactions in solution. The only difference is that, in solution reactions, you frequently need to convert concentrations to numbers of moles (or vice versa).

Typical calculations of this type are shown in Example 18.1. Problems 18.2, 18.9, and 18.28 are analogous to the example.

3. **Given the formula of an ionic compound, write the expression for K_{sp}.**

The principle is the same as that introduced in Chapter 15 in connection with K_c. Thus we have:

$$AgCl(s) \rightleftharpoons Ag^+(aq) + Cl^-(aq); \quad K_{sp} = [Ag^+][Cl^-]$$

$$PbCl_2(s) \rightleftharpoons Pb^{2+}(aq) + 2\,Cl^-(aq); \quad K_{sp} = [Pb^{2+}][Cl^-]^2$$

$$FeF_3(s) \rightleftharpoons Fe^{3+}(aq) + 3\,F^-(aq); \quad K_{sp} = [Fe^{3+}][F^-]^3$$

$$As_2S_3(s) \rightleftharpoons 2\,As^{3+}(aq) + 3\,S^{2-}(aq); \quad K_{sp} = [As^{3+}]^2[S^{2-}]^3$$

Note that here, as always, solids are omitted from the equilibrium constant expression.

4. **Given the solubility (mol/dm^3) of an ionic compound, calculate K_{sp}. Carry out the reverse calculation.**

Both of these calculations are shown in Example 18.3. Note that the relationship between these two quantities depends upon the type of salt, as shown in the following table (where S = solubility).

Type of Salt	MX	MX$_2$ or M$_2$X	MX$_3$ or M$_3$X	M$_2$X$_3$ or M$_3$X$_2$
K$_{sp}$	S^2	4S^3	27S^4	108S^5
Example	AgCl	PbCl$_2$, Ag$_2$S	FeF$_3$	As$_2$S$_3$

The conversion from solubility to K$_{sp}$ (and from grams to moles) is required in Problems 18.14 and 18.33. The reverse calculation is involved in Problems 18.13 and 18.32. Problem 18.3 illustrates both types of calculations.

5. Use the value of K$_{sp}$ to:

   a. determine the concentration of an ion in solution, given that of the other ion in equilibrium with it.

The solubility product of PbCl$_2$ is 1.7×10^{-5}. What is the concentration of Pb^{2+} in equilibrium with 0.020 M Cl$^-$? _____
Substituting directly into the expression for K$_{sp}$:

$$1.7 \times 10^{-5} = [\text{Pb}^{2+}][\text{Cl}^-]^2 = [\text{Pb}^{2+}](2.0 \times 10^{-2})^2$$

Solving:

$$[\text{Pb}^{2+}] = \frac{1.7 \times 10^{-5}}{(2.0 \times 10^{-2})^2} = \frac{1.7 \times 10^{-5}}{4.0 \times 10^{-4}} = 0.42 \times 10^{-1} = 0.042 \text{ M}$$

This skill is often involved as one step in a more complex problem. Consider, for instance, Example 18.4, in which the ultimate aim is to obtain the percentage of Mg^{2+} left in seawater after adding OH$^-$; Problems 18.4(a), 18.16, and 18.35 are similar.

   b. decide whether or not a precipitate will form when two solutions are mixed.

The principle here is a simple one: a precipitate will form only if the concentration product exceeds K$_{sp}$. Otherwise, the solution will be unsaturated with respect to the solid, and solubility equilibrium will not be established. The skill is illustrated in Example 18.5 and Problems 18.4(b),

244 • 18–Precipitation Reactions

18.15, 18.17, 18.34, and 18.36. You may notice a common assumption ordinarily made in problems of this type: when two solutions are mixed it is assumed that there is no net change in volume (i.e., 400 cm³ of solution 1 mixed with 800 cm³ of solution 2 gives a final volume of 1200 cm³).

    c. decide which of two possible precipitates will form when two solutions are mixed.

This skill is involved in the first part of Problems 18.18 and 18.37. The general principle is that the precipitate that requires the lowest concentration of reagent will form first. Thus, in Problem 18.18, we calculate that the concentrations of S^{2-} required to start to precipitate CoS and FeS are:

CoS: $[S^{2-}] = K_{sp}$ CoS/$[Co^{2+}]$ = $1 \times 10^{-21}/1 \times 10^{-1}$ = 1×10^{-20}

FeS: $[S^{2-}] = K_{sp}$ FeS/$[Fe^{2+}]$ = $1 \times 10^{-18}/1 \times 10^{-1}$ = 1×10^{-17}

Since $1 \times 10^{-20} < 1 \times 10^{-17}$, we deduce that CoS precipitates first.

    6. **Calculate the percentage of a species present in a mixture, given analytical data for a precipitation reaction.**

Example 18.6 illustrates a typical calculation of gravimetric analysis, in which a weighed sample is converted to a weighed precipitate. Here, all that is required is to establish the conversion factor which relates the mass of precipitate (As_2S_3) to the mass of the substance being analyzed for (As). Once this factor (245.8 g $As_2S_3 \triangleq$ 149.8 g As) is established, the problem unravels. Indeed, it becomes just another example of the mass-mass conversions first introduced in Chapter 3. See Problems 18.5, 18.20, and 18.39.

Example 18.7 is typical of calculations required in volumetric analysis. Here one measures the volume of a reagent of known concentration required to react with a sample of known mass. The first step is to calculate the number of moles of reagent used ($AgNO_3$) and then relate that to the number of moles of the species being analyzed for (Cl^-). Then one calculates the number of grams of that species, and finally its percentage in the sample. Problems 18.21 and 18.40 are entirely analogous to Example 18.7.

    7. **Use a table of solubilities to develop a scheme for separating a mixture of ions.**

Refer to Example 18.8 and the discussion immediately following. See also Problems 18.22, 18.23, 18.41, and 18.42.

8. Use precipitation reactions to "convert" one ionic compound to another.

This skill is described in the text. The basic idea is a simple one. To convert a compound MX (e.g., CsCl) to another compound MY (e.g., $CsNO_3$), we add a reagent ($AgNO_3$) containing the desired anion Y and a cation which forms a precipitate with the anion we want to get rid of, X. By filtering off this precipitate, we effectively substitute Y for X.

What reagent should be added to a solution of Na_2SO_4 to give, after filtration, a solution of NaCl?_____
Clearly the anion of the reagent must be Cl^-. The cation should be one that gives a precipitate with SO_4^{2-}. A suitable choice would be Ba^{2+}, since $BaSO_4$ is insoluble. In other words, we would add a solution of $BaCl_2$ and filter off the $BaSO_4$ to obtain a solution of NaCl.

See Problems 18.6, 18.25, and 18.44.

SELF-TEST

True or False

1. Since F^- is smaller than Cl^-, you would expect CaF_2 to be ( ) less soluble than $CaCl_2$.

2. Anion-anion contact in $MgSO_4$ should make it less soluble ( ) than $BaSO_4$.

3. The solubility of AgCl in 0.10 M NaCl is greater than it is in ( ) water.

4. The solubility of AgCl in 0.10 M $NaNO_3$ is greater than it is ( ) in water.

5. Any two salts that have the same K_{sp} value will have the ( ) same solubility.

6. When equal volumes of 0.1 M solutions of M^+ and X^- are ( ) mixed, a precipitate forms if K_{sp} of MX is less than 0.01.

7. A precipitate will form when enough Cl^- is added to a ( ) solution 0.10 M in Pb^{2+} to make conc. $Cl^- = 0.010$ M (K_{sp} $PbCl_2 = 1.7 \times 10^{-5}$).

8. A salt, MNO_3, can be prepared by a precipitation reaction ( ) which involves adding $AgNO_3$ to a solution of MCl.

9. The observed solubility of an ionic compound is usually ( ) greater than that calculated from the K_{sp}.

10. A saturated solution of $CaSO_4$ might well contain an ( ) appreciable concentration of ion pairs.

Multiple Choice

11. When 200 cm³ of 0.10 M $BaCl_2$ is added to 100 cm³ of ( ) 0.30 M Na_2SO_4, the number of moles of $BaSO_4$ precipitated is
    (a) 0.010   (b) 0.020
    (c) 0.030   (d) 0.20

12. 200 cm³ of 0.10 M $NiCl_2$ is added to 100 cm³ of 0.20 M ( ) NaOH. The number of moles of Ni^{2+} left in solution after precipitation is
    (a) 0   (b) 0.010
    (c) 0.030   (d) 0.10

13. When solutions of $Pb(NO_3)_2$ and Na_2SO_4 are mixed, the ( ) precipitate that forms is
    (a) $PbSO_4$ (b) $NaNO_3$
    (c) $PbSO_4$ and $NaNO_3$ (d) none

14. The net ionic equation for the reaction that occurs when ( ) solutions of $AgNO_3$ and Na_2CO_3 are mixed is
    (a) $2 AgNO_3(aq) + Na_2CO_3(aq) \rightarrow Ag_2CO_3(s) + 2 NaNO_3(aq)$
    (b) $2 Ag^+(aq) + CO_3^{2-}(aq) \rightarrow Ag_2CO_3(s)$
    (c) $Na^+(aq) + NO_3^-(aq) \rightarrow NaNO_3(s)$
    (d) no reaction

15. The solubility product of $PbCO_3$ is 1×10^{-12}. In a solution ( ) in which $[CO_3^{2-}] = 0.2$ M, the equilibrium concentration of Pb^{2+} is
    (a) 1×10^{-12} M   (b) 5×10^{-12} M
    (c) 2×10^{-11} M   (d) 1×10^{-6} M

16. The solubility product expression for As_2S_3 is: $K_{sp} =$ ( )
    (a) $[As_2^{3+}][S_3^{2-}]$ (b) $[As^{3+}]^2[S^{2-}]$
    (c) $[As^{3+}]^3[S^{2-}]^2$ (d) none of these

17. In order to remove 90% of the Ag^+ from a solution ( )
originally 0.10 M in Ag^+, the $[CrO_4^{2-}]$ must be (K_{sp} $Ag_2CrO_4 = 2 \times 10^{-12}$)
    (a) 2×10^{-12}
    (b) 2×10^{-11}
    (c) 2×10^{-10}
    (d) 2×10^{-8}

18. For a salt of formula MX_2, the solubility, S, will be related ( )
to K_{sp} by
    (a) $S = K_{sp}$
    (b) $S^2 = K_{sp}$
    (c) $2 S^3 = K_{sp}$
    (d) $4 S^3 = K_{sp}$

19. CrO_4^{2-} is used as an indicator in the titration of Cl^- with ( )
Ag^+ because
    (a) it is yellow, whereas Cl^- is colorless
    (b) Ag_2CrO_4, unlike AgCl, is soluble in water
    (c) Ag_2CrO_4 precipitates before AgCl
    (d) Ag_2CrO_4 precipitates only when virtually all the Cl^- has reacted

20. To separate and identify the ions in a mixture that may ( )
contain Pb^{2+}, Cu^{2+}, and Mg^{2+}, one might add the reagents H_2S, HCl, and NaOH. They should be added in the order
    (a) HCl, H_2S, NaOH
    (b) H_2S, HCl, NaOH
    (c) HCl, NaOH, H_2S
    (d) NaOH, H_2S, HCl

21. To prepare pure RbCl from Rb_2SO_4, one might use a ( )
precipitation reaction in which the other reagent is
    (a) $BaCl_2$
    (b) $BaSO_4$
    (c) NaCl
    (d) RbCl

22. Consider the anions Cl^-, SO_4^{2-}, CO_3^{2-}.
    (a) Which form insoluble salts with Pb^{2+}?_____
    (b) Which form insoluble salts with Ba^{2+}?_____
    (c) Which decrease the water solubility of KOH?_____
    (d) Which decrease the water solubility of $BaSO_4$?_____

23. The addition of $AgNO_3$ to a saturated solution of AgCl ( )
would
    (a) cause more AgCl to precipitate
    (b) increase the solubility of AgCl due to the interionic attraction of NO_3^- and Ag^+
    (c) lower the value of K_{sp} for AgCl
    (d) shift to the right the equilibrium $AgCl(s) \rightleftarrows Ag^+(aq) + Cl^-(aq)$

248 • 18–Precipitation Reactions

24. Sodium chloride is soluble in water; yet, when concentrated ( )
hydrochloric acid is added to a saturated solution of this salt, NaCl(s) precipitates out. Why?
    (a) HCl is a strong acid, and any strong acid will cause the precipitation.
    (b) The common ion, Cl^-, shifts the equilibrium; NaCl(s) forms so that $[Na^+][Cl^-]$ remains constant.
    (c) The K_{sp} decreases in the presence of acid.
    (d) The K_{sp} is unaffected by acid, but is reduced by the increase in $[Cl^-]$.

25. In a saturated solution of $Cr(OH)_3$ the concentration of ( )
OH^- was found to be 6.3×10^{-8} M. The solubility of $Cr(OH)_3$ (mol/dm^3) is
    (a) 6.3×10^{-8} (b) 2.1×10^{-8}
    (c) 4.0×10^{-15} (d) 4.3×10^{-28}

26. $CaSO_4$ has a $K_{sp} = 3 \times 10^{-5}$. In which of the following ( )
should $CaSO_4$ be most soluble?
    (a) 1 M H_2SO_4 (b) 2 M $CaCl_2$
    (c) pure H_2O (d) same in all three

27. If 1 M solutions of the following pairs of substances are ( )
mixed, in which case would a precipitate form?
    (a) $(NH_4)_2CO_3 + ZnCl_2$ (b) $Na_2S + Sr(NO_3)_2$
    (c) $NiCl_2 + MgSO_4$ (d) $KCl + Ba(OH)_2$

28. The volume of 0.10 M $AgNO_3$ that is required to react ( )
exactly with 100 cm^3 of 0.10 M Na_2S to form a precipitate of silver sulfide is
    (a) 100 cm^3 (b) 2.0 dm^3
    (c) 200 cm^3 (d) 20 cm^3

29. Suppose you have a solution containing both Ca^{2+} and Ag^+ ( )
at 0.001 M. You wish to separate the two ions by precipitation of a salt of one of them. After referring to the table of K_{sp} values choose the most suitable reagent from the list below:
    (a) 0.1 M $C_2H_3O_2^-$ (b) 0.1 M NO_3^-
    (c) 0.1 M I^- (d) H_2O

30. Which of the following can be correctly called a net ionic ( )
equation?
    (a) $Ni^{2+}(aq) + 2 Cl^-(aq) + 2 OH^-(aq) \to Ni(OH)_2(s) + 2 Cl^-(aq)$
    (b) $Ba^{2+}(aq) + 2 Br^-(aq) + 2 Na^+(aq) + SO_4^{2-}(aq) \to BaSO_4(s) + 2 Na^+(aq) + 2 Br^-(aq)$
    (c) $Ag^+(aq) + I^-(aq) \to AgI(s)$
    (d) both (a) and (b)

31. The solubility of PbCl$_2$ is 0.99 g/100 g H$_2$O at 20°C and ( ) 3.34 g/100 g H$_2$O at 100°C. For the reaction PbCl$_2$(s) → Pb^{2+}(aq) + 2 Cl$^-$(aq), ΔH
    (a) is positive   (b) is negative
    (c) equals zero   (d) cannot be determined

Problems

32. A solid sample contains MgSO$_4$ and NaCl. Treating an aqueous solution of 0.750 g of this sample with BaCl$_2$ solution yields 0.930 g of BaSO$_4$. What is the percentage of MgSO$_4$ in the original sample? (AM Mg = 24.3, Ba = 137, S = 32.1, O = 16.0)

33. The solubility of PbCl$_2$ in water at a certain temperature is 1.6 × 10^{-2} mol/dm^3.
    (a) Calculate the solubility product constant, K$_{sp}$, for PbCl$_2$.
    (b) What is the equilibrium concentration of Pb^{2+} when PbCl$_2$(s) is mixed with 2.0 M NaCl?

34. In a saturated solution of Zn(OH)$_2$, the OH$^-$ concentration is 4.0 × 10^{-6} M. Calculate
    (a) K$_{sp}$ of Zn(OH)$_2$
    (b) the solubility of Zn(OH)$_2$, FM = 100, (g/100 cm^3).

*35. A 0.500-g mixture of KCl and KBr is dissolved in water and treated with excess AgNO$_3$ solution. The precipitate which forms is found to have a mass of 0.861 g. What percentage of the mixture is KCl?

SELF-TEST ANSWERS

1. T (High charge density results in stronger attractions in the solid; see Chapters 9, 12.)
2. F (Rather, anion-anion contact reduces attractions in the solid, making it more soluble.)
3. F (Common ion effect.)
4. T (Ion "atmospheres" reduce ion-ion repulsions, enhancing solubility — Chapter 13.)
5. F (Only if they have the same type formula, e.g., MX and AB.)
6. F (Dilution lowers conc. to 0.05 M, so K$_{sp}$ would have to be < (0.05)^2.)
7. F (1 × 10^{-5} < 1.7 × 10^{-5}.)
8. T
9. T

250 • 18–Precipitation Reactions

10. T (See the text discussion in Chapters 13 and 18.)
11. b (Ba^{2+} is the limiting reactant; how many moles are there of Ba^{2+} and SO_4^{2-}? Such calculations were done in Chapter 12.)
12. b (0.020 mol Ni^{2+} and of OH^- initially present; they react 1 mol to 2 mol.)
13. a (The only insoluble combination of cation and anion.)
14. b (Representing all that could be observed.)
15. b ($K_{sp}/[CO_3^{2-}]$.)
16. d ($[As^{3+}]^2[S^{2-}]^3$.)
17. d (Note that $[Ag^+] = 0.010$, so $[CrO_4^{2-}] = K_{sp}/[Ag^+]^2 = K_{sp}/(0.010)^2$.)
18. d ($[M] = S$ while $[X] = 2S$.)
19. d (Indicators reveal the "completion" of reaction, i.e., approaching 100%.)
20. a (Any other order would precipitate two ions at once.)
21. a
22. (a) all   (b) SO_4^{2-}, CO_3^{2-} (c) none   (d) SO_4^{2-}
23. a (Common ion effect.)
24. b (For the saturated solution, we have $NaCl(s) \rightleftarrows Na^+(aq) + Cl^-(aq)$.)
25. b (Here, $[OH^-] = 3[Cr^{3+}] = 3 \times$ solubility.)
26. c (Why?)
27. a (Check the table of solubility rules; you get $ZnCO_3$.)
28. c (The precipitate must have the formula Ag_2S. To recognize charges on ions, see Chapter 8. Required: 0.020 mol Ag^+. For calculations involving concentrations, see Chapter 12.)
29. c (Precipitates "all" Ag^+.)
30. c
31. a (Equilibrium lies farther to the right at higher T for endothermic reactions. Use Le Chatelier's Principle – Chapter 15.)
32. $0.930 \text{ g } BaSO_4 \times \dfrac{120 \text{ g } MgSO_4}{233 \text{ g } BaSO_4} = 0.479 \text{ g } MgSO_4$

$\% \text{ } MgSO_4 = \dfrac{0.479}{0.750} \times 100 = 63.9\%$

33. (a) $K_{sp} = (1.6 \times 10^{-2})(3.2 \times 10^{-2})^2 = 1.6 \times 10^{-5}$
    (b) $[Pb^{2+}] = 1.6 \times 10^{-5}/(2.0)^2 = 4.0 \times 10^{-6}$
34. (a) $[OH^-] = 4.0 \times 10^{-6}$, $[Zn^{2+}] = 2.0 \times 10^{-6}$
    $K_{sp} = (2.0 \times 10^{-6})(4.0 \times 10^{-6})^2 = 3.2 \times 10^{-17}$
    (b) $2.0 \times 10^{-6} \dfrac{\text{mol}}{\text{dm}^3} \times 100 \dfrac{\text{g}}{\text{mol}} = 2.0 \times 10^{-4} \text{ g/dm}^3 = 2.0 \times 10^{-5} \text{ g/100 cm}^3$

*35. The precipitate contains both AgCl and AgBr. (Except for recognizing this, the problem could just as well have been written for Chapter 3.)

$$0.861 \text{ g} = \text{moles AgBr}(188 \text{ g/mol}) + \text{moles AgCl }(143 \text{ g/mol})$$

or, $0.861 = \text{moles KBr }(188) + \text{moles KCl }(143)$

moles KBr = grams KBr (1 mol/119 g);
moles KCl = grams KCl (1 mol/74.6 g)

$$\text{grams KBr} = 0.500 - \text{grams KCl}$$

Substituting:

$$0.861 = \frac{(0.500 - X)}{119}(188) + \frac{X}{74.6}(143), \text{ where } X = \text{gram KCl}$$

$$X = 0.21$$

% KCl = 42

SELECTED READINGS

For practice in working solubility equilibria problems, see the manuals listed in the Preface.

Rather few articles can be found by this author that deal with the topics of this chapter. See what you can find. Meanwhile, consider the nature of molecular activity during the formation of a crystalline precipitate:

Fullman, R. L., The Growth of Crystals, *Scientific American* (March 1955), pp. 74–80.

19
ACIDS AND BASES

QUESTIONS TO GUIDE YOUR STUDY

1. What properties are characteristic of an acid? A base? Can you think of some common examples of each, other than "stomach acid"?

2. How are these properties explained in terms of atomic-molecular structure?

3. What happens when a substance dissolves in water (reacts with it?) to give an acidic solution? What happens when a solution forms which acts as a base? What species are present in such solutions?

4. Can you predict what species will give basic solutions; acidic solutions?

5. What is meant by the *strength* of an acid? How are differences in acid strength explained in terms of atomic-molecular structure? (Likewise for bases?)

6. How are such comparisons made quantitative?

7. How do you quantitatively describe a water solution of an acid or a base? (Recall that you have been able to specify the relative concentrations of species in gaseous systems at equilibrium and in saturated water solutions of slightly soluble salts.)

8. How would you experimentally arrive at such a description?

9. What role does water play in the formation of acidic and basic solutions? Is water necessary for the existence of an acid or base?

10. Why study acids and bases? Where do you encounter them?
11.

12.

254 • 19–Acids and Bases

YOU WILL NEED TO KNOW

Concepts

1. How to write Lewis structures — Chapter 8. Other concepts, primarily dealing with chemical equilibrium (Chapter 15), are used here mostly in quantitative applications, as described below.

Math

1. How to find logs and antilogs — Appendix 4.
2. How to perform stoichiometric calculations, particularly those involving concentration units — Chapters 3, 12 (and more worked examples in Chapter 18).
3. How to write the equilibrium constant expression for any reaction — Chapters 15, 18.
4. How to qualitatively and quantitatively predict the effects of changes in reactant or product concentrations on the position of equilibrium — Chapter 15.
5. How to use the Rule of Multiple Equilibria — Chapter 15.

CHAPTER SUMMARY

This chapter and Chapter 20 deal with the general topic of acid-base reactions. In this chapter we consider the properties of acids and bases; in Chapter 20 the reactions between them are discussed.

An acidic water solution is one in which there is an excess of H^+ ions; i.e., the concentration of H^+ is greater than that of OH^-. Since, for any water solution at 25°C, $[H^+] \times [OH^-] = 1.0 \times 10^{-14}$, it follows that in an acidic solution, $[H^+] > 10^{-7}$ M. By the same token, a basic water solution is one in which there are excess OH^- ions; i.e., $[OH^-] > [H^+]$ or $[OH^-] > 10^{-7}$ M. Acidity or basicity can also be expressed in terms of pH, defined by the equation $pH = -\log_{10} [H^+]$. A solution of pH 7 is said to be neutral; an acidic solution has a pH less than 7 while a basic solution has a pH greater than 7.

Any species which forms H^+ ions when added to water will form an acidic solution. From this point of view, the following would be classified as acids:

1. A large number of molecules containing ionizable hydrogen atoms. These may be binary compounds (e.g., $HCl \rightarrow H^+ + Cl^-$) or oxyacids

(e.g., $HNO_3 \rightarrow H^+ + NO_3^-$), in which the ionizable hydrogen atom is covalently bonded to oxygen.
2. A few anions containing ionizable hydrogen atoms, of which the HSO_4^- ion is typical: $HSO_4^- \rightarrow H^+ + SO_4^{2-}$.
3. Many, indeed most, cations. A hydrated cation such as $Zn(H_2O)_4^{2+}$ can lose a proton to give an acidic solution ($Zn(H_2O)_4^{2+} \rightarrow Zn(H_2O)_3OH^+ + H^+$). The only cations which show no tendency to react in this way are those of the 1A metals and the large ions of Group 2A.

Species which form basic solutions when added to water include:
1. The hydroxides of the 1A and 2A metals ($NaOH(s) \rightarrow Na^+ + OH^-$).
2. Ammonia ($NH_3 + H_2O \rightarrow NH_4^+ + OH^-$) and its organic derivatives such as CH_3NH_2.
3. Many, indeed most, anions, of which the F^- ion is typical:

$$F^- + H_2O \rightarrow HF + OH^-$$

A relatively few *strong acids* ionize completely when added to water. These include HCl, HBr, HI, HNO_3, $HClO_4$, and H_2SO_4 (1st ionization). Species which ionize only partially, called *weak acids*, are much more common. The strength of an acid is directly related to the magnitude of its equilibrium constant for ionization, K_a:

$$HX(aq) \rightleftharpoons H^+(aq) + X^-(aq); K_a = \frac{[H^+][X^-]}{[HX]}$$

The only strong bases are the hydroxides of the 1A and 2A metals, which are virtually completely ionized in water. Ammonia and its organic derivatives are weak bases, in the sense that the reaction

$$NH_3(aq) + H_2O \rightleftharpoons NH_4^+(aq) + OH^-(aq); K_b = \frac{[NH_4^+][OH^-]}{[NH_3]}$$

is incomplete. Anions derived from weak acids (e.g., F^-, CN^-, $C_2H_3O_2^-$, CO_3^{2-}) act as weak bases in water solution. The dissociation constant, K_b, of any weak base can be calculated by applying the Multiple Equilibrium Rule to show that

$$K_b \times K_a = K_w = 1.0 \times 10^{-14}$$

where K_a is the dissociation constant of the corresponding acid, often called the conjugate weak acid.

The relative strengths of related acids can in fact be estimated from structural considerations. Among a series of acids derived from the same element (e.g., $HClO_4$,

HClO$_3$, HClO$_2$, HClO), acid strength increases with the number of oxygen atoms attached to the central atom. Among acids of the same type formula derived from different nonmetals (e.g., H$_2$SO$_3$, H$_2$SeO$_3$, H$_2$TeO$_3$), acid strength increases with the electronegativity of the central atom. Increasing acid strength results from decreasing the H—O bond strength by the removal of electrons from that bond.

Throughout most of this chapter, we have, at least by implication, used the Arrhenius definition of acids and bases as species which upon addition to water give H$^+$ or OH$^-$ ions, respectively. The Brönsted-Lowry definition is somewhat broader; here an acid is taken to be a proton donor while a base is a proton acceptor. Thus, for the reaction:

$$NH_3(aq) + H_2O \rightleftharpoons NH_4^+(aq) + OH^-(aq)$$

the species NH$_3$ and OH$^-$ are acting as Brönsted-Lowry bases while H$_2$O and NH$_4^+$ are acids. A still more general definition due to G. N. Lewis takes an acid to be an electron-pair acceptor and a base an electron-pair donor. Metal cations such as Zn^{2+} act as Lewis acids when they accept pairs of electrons from H$_2$O or NH$_3$ molecules to form the complex ions Zn(H$_2$O)$_4$$^{2+}$ and Zn(NH$_3$)$_4$$^{2+}$ (Chapter 21).

BASIC SKILLS

Students often find the material in this chapter rather difficult to master, principally because a large number of different concepts are introduced. Some of these are quantitative in nature; others are qualitative. These are grouped separately below.

Quantitative Skills

1. Given one of the three quantities for a water solution, [H$^+$], [OH$^-$], and pH, calculate the other two quantities.

The basic relationships are:

$$pH = -\log_{10} [H^+]; [H^+][OH^-] = 1.0 \times 10^{-14}$$

The use of these relations is shown in Example 19.1. They are applied in Problems 19.1, 19.9, 19.10, 19.11, 19.29, 19.30, and 19.31. You can assume that all the acids and bases listed in Problems 19.11 and 19.31 are "strong," i.e., completely dissociated in water.

2. Calculate K$_a$ for a weak acid HA, given the [H$^+$] or pH of a solution prepared by dissolving HA in water to give a known initial concentration of HA.

This calculation is illustrated in Example 19.3. Note the two basic relationships which must apply when the solution is formed simply by dissolving HA in water:

$$[H^+] = [A^-]; [HA] = \text{orig. conc. HA} - [H^+]$$

Problems 19.2, 19.14, and 19.34 are similar to this example, except that you first have to calculate $[H^+]$ from the pH given.

3. **Given the original concentration of a weak acid and the value of K_a, calculate $[H^+]$ or per cent dissociation.**

The calculation of $[H^+]$ is discussed in considerable detail in Examples 19.4 and 19.5. In the first example, the concentration of H^+ is so small compared to the original concentration of weak acid that the approximation

$$[HA] = \text{orig. conc. HA} - [H^+] \approx \text{orig. conc. HA}$$

is entirely justified. In contrast, in Example 19.5 the $[H^+]$ calculated is an appreciable fraction (9.2%) of the original concentration of weak acid, too large to be ignored. For that reason, it is desirable to refine the calculation, using a second approximation more nearly valid than the first.

The per cent dissociation of a weak acid is readily calculated from the defining relation:

$$\% \text{ diss.} = \frac{[H^+]}{\text{orig. conc. HA}}$$

You should be able to demonstrate for yourself that the per cent dissociation is 0.42% in Example 19.4 and 8.7% in Example 19.5

Problems 19.3, 19.15(b), 19.16, 19.35(b), and 19.36 illustrate this skill; the acids in Problems 19.15(a) and 19.35(a) are strong and hence 100% dissociated.

4. **Relate the value of K_a for a weak acid to K_b for its conjugate base.**

The relation here is:

$$K_b \text{ of } A^- = (1.0 \times 10^{-14})/K_a \text{ of HA}$$

See Example 19.7 and Problems 19.4, 19.18, and 19.38.

5. **Given the original concentration of a weak base and the value of K_b, calculate $[OH^-]$ and the per cent dissociation.**

258 • 19–Acids and Bases

The calculations here are entirely analogous to those for weak acids. Compare Example 19.6 to Example 19.4, Problems 19.17(b) and 19.15(b), 19.37(b) and 19.35(b).

Qualitative Skills

6. **Predict whether a given acid or base is strong or weak.**

This prediction is called for directly in Problem 19.5 It is involved indirectly in several problems (e.g., 19.15, 19.25, 19.35, 19.45). There are only six common strong acids: HCl, HBr, HI, HNO_3, $HClO_4$ and H_2SO_4. All other acids that you encounter in this course can be assumed to be weak. The strong bases are the hydroxides of the 1A and 2A metals (e.g., NaOH, $Ca(OH)_2$); all other bases can be assumed to be weak.

7. **Predict whether a particular ion will be neutral, acidic, or basic in water solution.**

Predictions of this sort can be made with the aid of Table 19.6. This table may seem formidable at first glance, but the principle behind it is really quite simple. The neutral anions are those derived from strong acids; the neutral cations are those derived from strong bases. Anions derived from weak acids are basic; cations derived from weak bases are acidic. There are essentially infinite numbers of ions in these two categories; those listed in the table as "basic anions" and "acidic cations" are only typical examples. Finally, there are a few oddball anions (HSO_4^- is one) which are acidic because they contain an ionizable proton.

This skill is involved in Problems 19.6, 19.19, 19.20, 19.21, 19.39, 19.40, and 19.41. In order to answer the parts of these problems that involve ionic compounds, you must first decide whether the individual ions are acidic, basic, or neutral.

8. **Predict whether a given ionic compound will give a neutral, basic, or acidic water solution.**

As Example 19.8 implies, there are four categories to be considered:

— neutral cation, neutral anion: neutral solution
— neutral cation, basic anion: basic solution
— acidic cation, neutral anion: acidic solution
— acidic cation, basic anion: may be acidic, basic, or neutral

This skill is involved in Problems 19.6, 19.19, and 19.39.

9. **Write a net ionic equation to explain why a species gives an acidic or a basic solution.**

Before attempting to write such an equation, you first have to decide whether the solution produced is acidic, basic, or neutral. If you decide it is neutral, there is no reaction to explain and no equation to write. If the solution is acidic, one product must be H^+. The other product is simply the conjugate base of the reactant. Examples include:

$$HF(aq) \rightarrow H^+(aq) + F^-(aq)$$

$$HSO_4^-(aq) \rightarrow H^+(aq) + SO_4^{2-}(aq)$$

$$NH_4^+(aq) \rightarrow H^+(aq) + NH_3(aq)$$

$$Zn(H_2O)_4^{2+}(aq) \rightarrow H^+(aq) + Zn(H_2O)_3(OH)^+(aq)$$

In order to make a solution basic, a species must ordinarily pick up a proton from a water molecule, leaving an OH^- ion behind. Thus, to explain the fact that solutions containing NH_3, F^-, and CO_3^{2-} are basic, we would write:

$$NH_3(aq) + H_2O \rightarrow NH_4^+(aq) + OH^-(aq)$$

$$F^-(aq) + H_2O \rightarrow HF(aq) + OH^-(aq)$$

$$CO_3^{2-}(aq) + H_2O \rightarrow HCO_3^-(aq) + OH^-(aq)$$

Following these examples, you should be able to write the equations called for in Problems 19.6, 19.20 and 19.40.

10. **Given the equation for an acid-base reaction, select the Brönsted acid and Brönsted base; the Lewis acid and Lewis base.**

This skill requires only that you apply the definition of Brönsted or Lewis acids and bases given in the text. See Problems 19.22 and 19.42. In Problems 19.24 and 19.44, you are required to identify potential Brönsted acids (species that are capable of giving up a proton), Brönsted bases (species that can accept a proton), Lewis acids (species that can accept an electron pair), and Lewis bases (species that can donate an electron pair). Here, it is helpful to start by writing the Lewis structure of the species. In the case of H_2O, for example, we see from its structure that it could lose a proton to form OH⁻, gain a proton to form H_3O^+, or donate a pair of unshared electrons. It could not, however, accept a pair of electrons, because there is no place to put them.

SELF-TEST

True or False

1. In a basic water solution at 25°C, $[H^+] > 10^{-7}$. ( )
2. It is impossible to have a solution with a negative pH. ( )
3. The pH of a solution prepared by dissolving 1×10^{-9} mol of HCl in 1000 cm³ of water is 9. ( )
4. A solution which gives off bubbles when Na_2CO_3 is added is acidic. ( )
5. A solution which is colorless to phenolphthalein (Table 19.3) must be acidic. ( )
6. There are more weak acids than strong acids. ( )
7. Solutions containing the CN⁻ ion are expected to be basic. ( )
8. HF ($K_a = 7 \times 10^{-4}$) is a stronger acid than HNO_2 ($K_a = 4 \times 10^{-4}$). ( )
9. A solution of NH_4Cl is basic. ( )
10. The conjugate anions of strong acids are strong bases. ( )

Multiple Choice

11. The concentration of H^+ in a solution is 2×10^{-4} M. The OH⁻ concentration is ( )
    (a) 2×10^{-4} M (b) 1×10^{-10} M
    (c) 2×10^{-10} M (d) 5×10^{-11} M

12. The pH of the solution in Question 11 is ( )
    (a) 3.0 (b) 3.7
    (c) 4.0 (d) 10.3

13. The solution referred to in Questions 11 and 12 ( )
    (a) is acidic (b) is basic
    (c) is neutral (d) cannot exist

14. The pH of a solution is 5.5. The concentration of H^+ is about ( )

(a) 1×10^{-6} M (b) 3×10^{-6} M
(c) 1×10^{-5} M (d) 3×10^{-5} M

15. A 0.10 M solution of HCl would have a pH of ( )
    (a) 0.00   (b) 1.00
    (c) 7.00   (d) 13.00

16. The pH of a 0.10 M solution of a weak acid would be ( )
    (a) less than 1   (b) 1
    (c) greater than 1   (d) cannot say

17. Which one of the following is *not* a strong acid? ( )
    (a) HCl   (b) HF
    (c) HNO$_3$ (d) HClO$_4$

18. Which one of the following is a strong base? ( )
    (a) Al(OH)$_3$ (b) NH$_3$
    (c) C$_2$H$_5$OH   (d) NaOH

19. Which one of the following ions would give a basic solution ( ) upon addition to water?
    (a) NH$_4^+$ (b) Na$^+$
    (c) C$_2$H$_3$O$_2^-$ (d) NO$_3^-$

20. A certain weak acid is 10% dissociated in 1.0 M solution. In ( ) 0.10 M solution, the percentage of dissociation would be
    (a) greater than 10   (b) 10
    (c) less than 10   (d) zero

21. K_b for NH$_3$ is 2×10^{-5}. K_a for the NH$_4^+$ ion is ( )
    (a) 2×10^{-5} (b) 5×10^{-9}
    (c) 5×10^{-10} (d) 2×10^{-19}

22. Of the following acids, which is the strongest? ( )
    (a) H$_2$TeO$_4$ (b) H$_2$SeO$_3$
    (c) H$_2$SeO$_4$ (d) H$_2$SO$_4$

23. In the reversible reaction HCO$_3^-$(aq) + OH$^-$(aq) ⇌ ( ) CO$_3^{2-}$(aq) + H$_2$O, the Brönsted acids are
    (a) HCO$_3^-$ and CO$_3^{2-}$ (b) HCO$_3^-$ and H$_2$O
    (c) OH$^-$ and H$_2$O   (d) OH$^-$ and CO$_3^{2-}$

24. In the reaction BF$_3$ + NH$_3$ → F$_3$B:NH$_3$, BF$_3$ accepts an ( ) electron pair and acts as
    (a) an Arrhenius base   (b) a Brönsted acid
    (c) a Lewis acid   (d) a Lewis base

25. For the reaction HPO$_4^{2-}$(aq) + H$_2$O → H$_2$PO$_4^-$(aq) + ( ) OH$^-$(aq)
    (a) HPO$_4^{2-}$ is an acid and OH$^-$ its conjugate base
    (b) H$_2$O is an acid and OH$^-$ its conjugate base

(c) HPO_4^{2-} is an acid and $H_2PO_4^-$ its conjugate base
(d) H_2O is an acid and HPO_4^{2-} its conjugate base

26. Consider the species Cu^{2+}, F^-, H_2O, NH_4^+, and SO_3.
   (a) Which, on addition to water, give acidic solutions?
   _____
   (b) Which, on addition to water, give basic solutions?
   _____
   (c) Which can act as Brönsted acids? _____
   (d) Which can act as Brönsted bases? _____
   (e) Which can act as Lewis acids? _____
   (f) Which can act as Lewis bases? _____

27. Which one of the following might you expect to be an ( ) "active ingredient" in Brand X Antacid?
   (a) KOH
   (b) $SO_2(OH)_2$
   (c) $NaHCO_3$
   (d) NH_4Cl

28. Consider two solutions, one at a pH of 2.0 and the other at ( ) a pH of 4.0. The $H^+(aq)$ concentrations compare like

$$\frac{[H^+] \ (pH = 2)}{[H^+] \ (pH = 4)} =$$

   (a) 1/2
   (b) log (1/2)
   (c) 10/1
   (d) 100/1

29. Which one of the following equations would you write to ( ) explain the fact that a water solution of NaCl is neutral?
   (a) $Na(H_2O)_4^+(aq) \rightarrow Na(H_2O)_3(OH)(aq) + H^+(aq)$
   (b) $Na(s) + H_2O \rightarrow Na^+(aq) + OH^-(aq) + ½ H_2(g)$
   (c) $Cl^-(aq) + H_2O \rightarrow HCl(aq) + OH^-(aq)$
   (d) none of the above

30. A strong acid in aqueous solution ( )
   (a) is a strong electrolyte
   (b) tastes sour
   (c) reacts with a weak base to give a salt whose solution is acidic
   (d) all the above

31. The K_a of an acid was found to be 1.8×10^{-9}. The ( ) electrical conductivity of a 1.0 M solution of this acid should be
   (a) very high
   (b) very low
   (c) similar to that of 1.0 M NaCl
   (d) similar to that of 1.0 M HNO_3

Problems

32. A bottle of the weak acid, aggravatic acid, labelled "0.040 M HA", is found to have a pH of 2.70. Calculate
    (a) [H$^+$] (b) [HA] (c) K_a for aggravatic acid

33. A solution of sodium acetate, $NaC_2H_3O_2$, is weakly basic.
    (a) Write a balanced net ionic equation to explain why it is basic.
    (b) Calculate K for the reaction in (a), given that K_a for acetic acid, $HC_2H_3O_2$, is 2×10^{-5}.
    (c) Calculate [OH$^-$] in a 1.00 M solution of $NaC_2H_3O_2$.

34. Consider the following substances:
    (a) LiOH (b) Al(NO$_3$)$_3$ (c) Na$_2$CO$_3$ (d) HF
    Predict whether water solutions of these compounds will be acidic, basic, or neutral, and write net ionic equations to explain your predictions.

*35. For HSO$_4^-$(aq), $K_a = 1.2 \times 10^{-2}$.
    (a) Calculate the concentration of H$^+$(aq) in 0.10 M KHSO$_4$.
    (b) Estimate the osmotic pressure of this solution at 25°C.

SELF-TEST ANSWERS

1. F ([OH$^-$] > 10^{-7}.)
2. F (e.g., 10 M HCl.)
3. F (Where did the OH$^-$ ions come from? See the Readings.)
4. T (Either that or it is supersaturated with gas! Write the equation for the former case.)
5. F (Could, for example, have a pH of 7.5.)
6. T (You should *know* the common strong acids and bases — Table 19.4.)
7. T (Its conjugate acid, HCN, is weak.)
8. T (Caution: direct comparison of K_a's requires similar formulas: e.g., HA and HB; the same number of H$^+$ ions per mole must be able to dissociate.)
9. F (NH$_4^+$ is acidic.)
10. F (Cl$^-$, for example, is neutral.)
11. d (K_w/[H$^+$].)
12. b (-log (2 × 10^{-4}).)
13. a ([H$^+$] > 10^{-7}.)
14. b (Or, 10^{-pH}.)
15. b
16. c (Incomplete dissociation, so [H$^+$] < 0.10.)

264 • 19–Acids and Bases

17. b
18. d (OH^- not readily formed from the alcohol, C_2H_5OH. Why?)
19. c (Predictable, if you remember the common *strong* acids and bases.)
20. a (% dissociation greater for more dilute solution, in order that $\dfrac{[H^+][A^-]}{[HA]}$ remains constant.)
21. c ($K_a K_b = K_w$ for conjugate acid-base pair.)
22. d (With the weakest O—H bond; a common strong acid.)
23. b (Proton donors.)
24. c
25. b (Conjugate pairs differ only by one proton; H_2O donates proton.)
26. (a) Cu^{2+}, NH_4^+, SO_3 (acid anhydride, Chapter 17); (b) F^-;
    (c) H_2O, NH_4^+;   (d) F^-, H_2O;   (e) Cu^{2+}, SO_3;
    (f) F^-, H_2O, SO_3
27. c ("Bicarbonate of soda." KOH, strong base, would be corrosive! $SO_2(OH)_2$ is a tricky but informative representation of sulfuric acid.)
28. d
29. d (Neither ion tends to accept H^+; neither can donate H^+.)
30. d (Example of c: $H^+(aq) + NH_3(aq) \rightarrow NH_4^+(aq)$.)
31. b (Very little dissociation. Conductivities were discussed in Chapter 13.)
32. (a) $\log[H^+] = -2.70 = 0.20 - 3.00$; $[H^+] = 2.0 \times 10^{-3}$
    (b) $[HA] = 0.040 - 0.002 = 0.038$ M
    (c) $K_a = \dfrac{(2.0 \times 10^{-3})^2}{0.038} = 1.1 \times 10^{-4}$
33. (a) $C_2H_3O_2^-(aq) + H_2O \rightarrow HC_2H_3O_2(aq) + OH^-(aq)$
    (b) $K = K_b = K_w/K_a = \dfrac{1 \times 10^{-14}}{2 \times 10^{-5}} = 5 \times 10^{-10}$
    (c) $\dfrac{[OH^-]^2}{1.0} = 5 \times 10^{-10}$; $[OH^-] = 2 \times 10^{-5}$
34. (a) basic: $LiOH(s) \rightarrow Li^+(aq) + OH^-(aq)$
    (b) acidic: $Al(H_2O)_6{}^{3+}(aq) \rightarrow Al(H_2O)_5(OH)^{2+}(aq) + H^+(aq)$
    (c) basic: $CO_3{}^{2-}(aq) + H_2O \rightarrow HCO_3^-(aq) + OH^-(aq)$
    (d) acidic: $HF(aq) \rightarrow H^+(aq) + F^-(aq)$
*35. (a) $\dfrac{[H^+]^2}{0.10 - [H^+]} = 1.2 \times 10^{-2}$

    Using quadratic formula: $[H^+] = 2.9 \times 10^{-2}$
    Or, use two successive approximations (% dissociation = 29%).

    (b) The original 0.10 M $KHSO_4$ gives 0.10 M K^+; and HSO_4^- dissociates to give 0.029 M H^+, 0.029 M $SO_4{}^{2-}$, and the remaining 0.071 M HSO_4^-. So, total concentration of ions is 0.229 M.

π = MRT, where M represents total concentration of solute particles in moles per cubic decimetre.

π = 0.229 (8.31) (298) = 570 kPa

SELECTED READINGS

Acid-base theory is discussed in a very readable series starting with:

Jensen, W. B., Lewis Acid-Base Theory, *Chemistry* (March 1974), pp. 11–14.

A couple of special topics:

Chilton, T. H., *Strong Water: Nitric Acid, Its Sources, Methods of Manufacture, and Uses*, Cambridge, Mass., MIT Press, 1968.

Mogul, P. H., Dilute Solutions of Strong Acids: the Effect of Water on pH, *Chemistry* (October 1969), pp. 14–17.

20
ACID-BASE REACTIONS

QUESTIONS TO GUIDE YOUR STUDY

1. What reactions involve the participation of an acid or a base, or both? What, for example, constitutes a *neutralization*?

2. What effect does acid or base strength have on the extent of these reactions?

3. What energy effects are associated with acid-base reactions? Can you explain them in terms of what the molecules and ions are doing?

4. How can you predict the spontaneity of an acid-base reaction?

5. How would you follow an acid-base reaction experimentally? What would you measure to determine, for example, its extent?

6. What can be said about the rates of acid-base reactions? About reactions in which an acid or a base plays the part of a catalyst?

7. What are the properties of a buffer system? How do you account for them?

8. Where do you find buffers? (Can you name one?) What are some applications?

9. What are some of the applications of acid-base reactions?

10. How does an acid-base indicator work?

11.

12.

YOU WILL NEED TO KNOW

Concepts

1. How the position of equilibrium is affected by a change in concentration of reactant or product — Chapter 15.
2. How to write net ionic equations — Chapter 18.
3. Most of the concepts introduced in the preceding chapter — for example, how to predict whether a given species will act as an acid or a base; how to recognize equations for which equilibrium constants are K_w, K_a, or K_b.

Math

1. How to use the Rule of Multiple Equilibria — Chapter 15.
2. How to work problems involving equilibrium constants in general — see Chapter 18 for solubility product constants; Chapter 19, for dissociation constants.
3. For other essential math, see this section of Chapter 19.

CHAPTER SUMMARY

The neutralization reaction

$$H^+(aq) + OH^-(aq) \rightarrow H_2O; \quad K = 1.0 \times 10^{14}, \text{ at } 25°C$$

takes place when any strong acid reacts with any strong base in water solution. If the acid is weak, it is more appropriate to write the equation as

$$HX(aq) + OH^-(aq) \rightarrow H_2O + X^-(aq)$$

since the principal species in solution before neutralization is the HX molecule rather than the H^+ ion. Equilibrium constants for the reactions of weak acids with strong bases are smaller than that for neutralization ($K = 1/K_b = K_a \times 10^{14}$) but are ordinarily large enough to drive the reaction virtually to completion. A similar situation exists in the reaction of a weak base with a strong acid where $K = 1/K_a = K_b \times 10^{14}$. Here, the weak base may be a molecule or ion in water solution (e.g., NH_3, CO_3^{2-}) or an anion of a water-insoluble salt:

$$BaCO_3(s) + 2\,H^+(aq) \rightarrow Ba^{2+}(aq) + H_2CO_3(aq)$$

(Since CO_2 is the volatile anhydride of H_2CO_3, the reaction may also be represented by

$$BaCO_3(s) + 2\,H^+(aq) \rightarrow Ba^{2+}(aq) + H_2O + CO_2(g).)$$

Reactions of this type are commonly used to bring salts containing the anions of weak or volatile acids (F^-, CO_3^{2-}, S^{2-}) into solution.

The concentration of an acid or base in water solution or in a solid mixture can be determined by titration, in which we measure the volume of a reagent of known concentration required to reach the equivalence point. The indicator used is ordinarily a weak acid. Ideally, we choose an indicator whose K_a is equal to the concentration of H^+ at the equivalence point. In a strong acid-strong base titration, the choice of indicator is not critical since $[H^+]$ changes rapidly near the equivalence point. For a weak acid or weak base, the pH changes much more slowly. Indeed, if an acid or base is very weak, the change in pH is so gradual that an ordinary acid-base titration cannot be carried out.

A solution containing a mixture of a weak acid and its conjugate weak base acts as a buffer in the sense that it prevents drastic change in pH when small amounts of strong acid or strong base are added. For a sodium acetate-acetic acid buffer, the reactions are

addition of strong acid: $H^+(aq) + C_2H_3O_2^-(aq) \rightarrow HC_2H_3O_2(aq)$

addition of strong base: $OH^-(aq) + HC_2H_3O_2(aq) \rightarrow C_2H_3O_2^-(aq) + H_2O$

Ideally, we use a buffer system with the K_a of the weak acid equal to the concentration of H^+ that we wish to maintain. Thus, we might use an $C_2H_3O_2^-$-$HC_2H_3O_2$ buffer ($K_a = 1.8 \times 10^{-5}$) to maintain pH in the range 4.5 to 5.0.

Acid-base reactions, like precipitation reactions discussed in Chapter 18, are widely used in chemical analysis and synthesis, often in commercial applications. Examples include:

a. Acid-base titrations to determine the per cent HCO_3^- in a mixture.
b. Separation of a mixture of $BaCO_3$ and $BaSO_4$ by adding a strong acid to bring CO_3^{2-} into solution.
c. Preparation of a volatile species such as CO_2 (add H^+ to $CaCO_3$) or NH_3 (add OH^- to a salt containing the NH_4^+ cation).
d. The Solvay process by which $NaHCO_3$ and Na_2CO_3 are produced, ultimately, from $CaCO_3$, $NaCl$, and H_2O.

270 • 20–Acid-Base Reactions

BASIC SKILLS

This chapter, like Chapter 19, covers a rather large number of skills. Here, also, it is helpful to classify these skills into two categories.

Quantitative Skills

1. Given appropriate data, calculate the equilibrium constant for an acid-base reaction.

This skill is illustrated in Examples 20.2 and 20.3; general expressions for K are given in Table 20.1. Example 20.4 shows how one can calculate K for a somewhat more complex reaction, that between an insoluble solid and a strong acid. To make such a calculation, it is perhaps simplest to use the Rule of Multiple Equilibria introduced in Chapter 15.

Problems 20.2, 20.9, 20.10, 20.30, and 20.31 offer you a chance to test this skill. One point to keep in mind, is that for all of the acid-base reactions considered in this chapter, the equilibrium constant is a very large number ($K \gg 1$). Consequently, these reactions can be expected to go virtually to completion under ordinary conditions.

2. Relate titration data for an acid-base reaction to:
   a. the concentration of an acid or a base in solution.

The calculations here are illustrated in Example 20.5. Problems 20.3, 20.13, and 20.34 are similar but note that, in titration with a base, one mole of H_2SO_4 furnishes two moles of H^+.

Problem 20.14 illustrates a slightly different method of establishing the concentration of a base (or acid) in solution, using a pure solid as a reactant. The calculations involved are very similar to those in Example 20.6, where a solid is identified from data obtained by titration with a basic solution of known concentration.

   b. the gram equivalent mass of an acid (i.e., the mass that reacts with one mole of OH^-).

- -

A sample of a solid acid weighing 0.150 g requires 24.0 cm³ of a solution 0.200 M in OH^- for complete reaction. What is the GEM of the acid?

We start by calculating the number of moles of OH⁻ used.

no. moles OH⁻ = (0.0240 dm³) (0.200 mol/dm³) = 4.80 × 10⁻³ mol OH⁻

Since 0.150 g of acid reacts with 4.80 × 10⁻³ mol OH⁻, the mass that reacts with one mole of OH⁻ must be:

$$\frac{1.50 \times 10^{-1} \text{ g acid}}{4.80 \times 10^{-3} \text{ mol OH}^-} = 0.312 \times 10^2 \frac{\text{g acid}}{\text{mol OH}^-} = 31.2 \text{ g acid/mol OH}^-$$

Hence, the GEM of the acid must be 31.2 g.

Problem 20.35 is entirely analogous to the example just worked.

3. **Given the equation for an acid-base reaction, relate N for a reagent to M, or its gram equivalent mass to its gram molecular mass.**

The general relations are

$$\text{GEM} = \text{GFM}/n; \quad N = n \times M$$

where, for an acid, n = no. moles OH⁻ reacting with one mole of acid

for a base, n = no. moles H⁺ reacting with one mole of base

See Problems 20.15 and 20.36.

4. **Given the ionization constant for an acid-base indicator, show by calculation what color it will have in a solution of known [H⁺].**

The color is determined by the ratio [In⁻]/[HIn]. If this ratio is large (≥10), the solution will have the characteristic color of the In⁻ ion. If it is small (≤0.1), the color of the HIn molecule will predominate. If the ratio is close to 1, the color observed will be intermediate between those of In⁻ and HIn.

The desired ratio is readily obtained by solving the expression for K_a:

$$K_a = \frac{[H^+][In^-]}{[HIn]}; \quad \frac{[In^-]}{[HIn]} = \frac{K_a}{[H^+]}$$

Thus, for an indicator with $K_a = 10^{-7}$, the ratio will have the values 0.1, 1, and 10, respectively, for solutions of pH 6, 7, and 8.

This skill is required in Problems 20.4, 20.16 and 20.37.

5. Given the composition of a buffer, determine its pH before and after the addition of known amounts of a strong acid or base.

Any calculation of this sort proceeds through three steps.

a. The ratio [HA]/[A$^-$] is first determined. For the original buffer, the amounts of HA and A$^-$ are either given or readily calculated from the statement of the problem.

Addition of x moles of a strong acid to the buffer increases the number of moles of HA by x and decreases the number of moles of A$^-$ by the same amount. Conversely, if x moles of a strong base are added, the number of moles of A$^-$ increases by x while that of HA decreases by x. One point to keep in mind: since HA and A$^-$ are present in the same solution, their concentrations must be in the same ratio as the numbers of moles, i.e.,

$$\frac{[HA]}{[A^-]} = \frac{\text{no. moles HA}}{\text{no. moles A}^-}$$

b. [H$^+$] is calculated from the equation for K_a, i.e.,

$$[H^+] = K_a \times \frac{[HA]}{[A^-]}$$

c. pH is obtained from the defining equation pH = $-\log_{10}$ [H$^+$].

Typical calculations are shown in Example 20.8. Problems 20.5, 20.20, and 20.41 are very similar, but note that in 20.20(d) the buffer has been destroyed by neutralizing all the formic acid. Problems 20.21 and 20.42 illustrate the application of the buffer principle to blood.

6. Given, or having calculated, the ratio [A$^-$]/[HA] in a solution and knowing [H$^+$], determine K_a for a weak acid.

This skill follows directly from the definition of K_a:

$$K_a = \frac{[H^+][A^-]}{[HA]}$$

Clearly, K_a is the product of [H$^+$] times the ratio [A$^-$]/[HA]. See Example 20.9 and Problems 20.22 and 20.43.

Qualitative Skills

7. Write net ionic equations to describe acid-base reactions.

The nature of the equation depends upon whether the acid and base are strong or weak, and whether both species are in water solution or one is a solid.

a. *Strong acid – strong base.* If both species are in solution, the equation is simply

$$H^+(aq) + OH^-(aq) \rightarrow H_2O$$

If the reactant is a solid base, the equation is written to reflect that fact. Thus, for the addition of a strong acid to solid magnesium hydroxide we would write

$$2\,H^+(aq) + Mg(OH)_2\,(s) \rightarrow 2\,H_2O + Mg^{2+}(aq)$$

b. *Weak acid – strong base.* For a reaction taking place in water solution, the reactants are the weak acid molecule or ion and the OH^- ion. The products are an H_2O molecule and the conjugate base of the weak acid. Typical examples include:

$$HC_2H_3O_2\,(aq) + OH^-(aq) \rightarrow H_2O + C_2H_3O_2^-\,(aq)$$

$$HCO_3^-(aq) + OH^-(aq) \rightarrow H_2O + CO_3^{2-}(aq)$$

$$NH_4^+(aq) + OH^-(aq) \rightarrow H_2O + NH_3\,(aq)$$

These are the equations that we would write to describe the reactions of a solution of NaOH with solutions of $HC_2H_3O_2$, $NaHCO_3$, and NH_4Cl, respectively.

c. *Strong acid – weak base.* In water solution the reactants are H^+ and the weak base (molecule or anion). The product is the conjugate acid of the weak base. For the reaction of a solution of any strong acid with solutions containing the NH_3 molecule, the S^{2-} ion or the CO_3^{2-} ion, we would write:

$$NH_3\,(aq) + H^+(aq) \rightarrow NH_4^+(aq)$$

$$S^{2-}(aq) + 2H^+(aq) \rightarrow H_2S(aq)$$

$$CO_3^{2-}(aq) + 2H^+(aq) \rightarrow H_2CO_3\,(aq) \text{ or } CO_2\,(aq) + H_2O$$

274 • 20–Acid-Base Reactions

Frequently in reactions of this type the "weak base" is an anion present as an insoluble solid (e.g., sulfide or carbonate). To represent the reaction of any strong acid with insoluble ZnS and $CaCO_3$ we would write:

$$ZnS(s) + 2\ H^+(aq) \rightarrow H_2S(aq) + Zn^{2+}(aq)$$

$$CaCO_3(s) + 2\ H^+(aq) \rightarrow H_2CO_3(aq) + Ca^{2+}(aq)$$

    Write balanced net ionic equations for the acid-base reactions that occur when:
    a. a solution of HCl is added to $Ca(OH)_2$ (s). _____
    b. a solution of HNO_2 is added to a solution of NaOH. _____
    c. a solution of HCl is added to $BaCO_3$ (s). _____

To write these equations, you must realize that HCl is a strong acid, entirely dissociated to H^+ and Cl^- in solution, while HNO_2 is a weak acid, present in solution primarily as the HNO_2 molecule.

$$2\ H^+(aq) + Ca(OH)_2(s) \rightarrow 2\ H_2O + Ca^{2+}(aq)$$

$$HNO_2(aq) + OH^-(aq) \rightarrow H_2O + NO_2^-(aq)$$

$$2\ H^+(aq) + BaCO_3(s) \rightarrow H_2O + CO_2(g) + Ba^{2+}(aq)$$

In Problems 20.1, 20.8, and 20.29, you have a chance to apply these principles in their simplest form. Many of the other problems require that you write net ionic equations as one step in the solution (e.g., Problems 20.11, 20.24, 20.32, and 20.45).

    8. **Given the K_a values of several indicators, choose one which is appropriate for a particular acid-base reaction.**

A general discussion of the principles involved is given in the text. Summarizing this discussion:

    a. If both acid and base are strong, almost any indicator will work.
    b. If the acid is strong and the base is weak, use an indicator with a $K_a > 10^{-7}$ (e.g., methyl red).
    c. If the acid is weak and the base is strong, use an indicator with a $K_a < 10^{-7}$ (e.g., phenolphthalein).
    d. If both acid and base are weak, forget it; no indicator will work.
See Problems 20.17 and 20.38.

9. **Use acid-base reactions to separate or distinguish between ions.**

This skill is required in Problems 20.6, 20.24, 20.25, 20.45, and 20.46 (in the last three problems, reactions other than the acid-base type are permitted). Two helpful hints:

a. Many acid-base reactions yield gases which are easily identified (CO_2, H_2S, SO_2, NH_3).
b. Insoluble solids containing a base anion (e.g., CO_3^{2-}, S^{2-}, OH^-) can ordinarily be brought into solution by treating with a strong acid. If the anion is neutral (e.g., SO_4^{2-}, Cl^-), addition of a strong acid will have no effect.

10. **Use acid-base reactions to prepare inorganic compounds.**

This skill is required in Problems 20.23 and 20.44. If you are baffled by one of the parts of these problems, refer back to the text where the principles are discussed. A couple of hints:

a. Many of the reactions which are used to separate or identify ions also serve to convert one compound to another. Thus, treatment of FeS(s) with hydrochloric acid gives off H_2S and leaves $FeCl_2$ in solution; after evaporation, you have, in effect, converted FeS to $FeCl_2$.
b. Simple neutralization is often a key step in a conversion. Obviously, NaOH could be converted to NaI by neutralization with HI followed by evaporation. In a less obvious case, you might convert $MgCl_2$ to MgI_2 by first precipitating $Mg(OH)_2$, then neutralizing the precipitate with HI, and finally evaporating the resulting solution.

SELF-TEST

True or False

1. The equilibrium constant for the reaction of a weak acid ( ) with a strong base is smaller than that for a strong acid-strong base reaction.

2. When 100 cm³ of 0.10 M $HC_2H_3O_2$ is added to 100 cm³ of ( ) 0.10 M NaOH, a neutral solution is produced.

3. When 100 cm³ of 0.10 M HCl is added to 100 cm³ of 0.10 ( ) M NaOH, a neutral solution of NaCl is formed.

4. The equilibrium constant for the reaction of $C_2H_3O_2^-$ with ( )
HCl is equal to $1/K_a$, where K_a is the dissociation constant for $HC_2H_3O_2$.

5. The gram equivalent mass of acetic acid, $HC_2H_3O_2$, in an ( )
acid-base titration, is always equal to its gram molecular mass.

6. At the equivalence point of the titration of a strong acid ( )
with a weak base, the pH is greater than 7.

7. A mixture of 100 cm³ of 1.0 M HCl with 100 cm³ of 2.0 M ( )
$NaC_2H_3O_2$ would act as a buffer.

8. A mixture of 100 cm³ of 1.0 M HCl with 100 cm³ of 1.0 ( )
M $NaC_2H_3O_2$ would act as a buffer.

9. Ammonia can be prepared by heating a solution of ( )
ammonium chloride with a strong acid.

10. Sodium hydrogen carbonate could be prepared by saturat- ( )
ing a solution of NaOH with CO_2 and evaporating.

Multiple Choice

11. The equation for the reaction of a water solution of the ( )
weak acid HF with a solution of NaOH is best written as
    (a) $H^+(aq) + F^-(aq) \to HF(aq)$
    (b) $H^+(aq) + OH^-(aq) \to H_2O$
    (c) $HF(aq) + OH^-(aq) \to H_2O + F^-(aq)$
    (d) $HF(aq) + NaOH(aq) \to H_2O + NaF(aq)$

12. The equation for the reaction of a water solution of ( )
ammonia with a water solution of HCl is best written as
    (a) $NH_3(aq) + H_2O \to NH_4^+(aq) + OH^-(aq)$
    (b) $NH_4^+(aq) \to NH_3(aq) + H^+(aq)$
    (c) $NH_3(aq) + H^+(aq) \to NH_4^+(aq)$
    (d) $NH_3(aq) + HCl(aq) \to NH_4^+(aq) + Cl^-(aq)$

13. The equation for the reaction of Ag_2CO_3, a water-insoluble ( )
solid, with a strong acid is best written as
    (a) $Ag_2CO_3(s) \rightleftarrows 2\,Ag^+(aq) + CO_3^{2-}(aq)$
    (b) $CO_3^{2-}(aq) + 2\,H^+(aq) \to CO_2(g) + H_2O$
    (c) $Ag_2CO_3(s) \to Ag_2O(s) + CO_2(g)$
    (d) $Ag_2CO_3(s) + 2\,H^+(aq) \to 2\,Ag^+(aq) + CO_2(g) + H_2O$

14. A certain weak acid has a dissociation constant of 1×10^{-4}. ( )
The equilibrium constant for its reaction with a strong base is:
    (a) 1×10^{-4} (b) 1×10^{-10}
    (c) 1×10^{10} (d) 1×10^{14}

15. When 500 cm³ of 0.10 M NaOH is reacted with 500 cm³ of ( )
0.20 M HCl, the final concentration of H⁺ is
   (a) 0.10 M   (b) 0.20 M
   (c) 0.050 M   (d) 10^{-7} M

16. Which one of the following salts will not be soluble in ( )
strong acid?
   (a) $PbCO_3$ (b) CaF_2
   (c) $PbSO_4$ (d) KCl

17. For the reaction $H_2PO_4^-(aq) + 2\ OH^-(aq) \rightarrow PO_4^{3-}(aq) + 2\ H_2O$, the gram equivalent mass of $H_2PO_4^-$ is equal to
   (a) 2 × GFM   (b) GFM
   (c) ½ GFM   (d) that of PO_4^{3-}

18. When Na_2CO_3 takes part in an acid-base reaction, the ratio ( )
of N to M is:
   (a) ½   (b) 1
   (c) either ½ or 1   (d) either 1 or 2

19. A certain buffer contains equal concentrations of X^- and ( )
HX. The K_b of X^- is 10^{-10}. The pH of the buffer is
   (a) 4   (b) 7
   (c) 10   (d) 14

20. A buffer is formed by adding 500 cm³ of 0.20 M $HC_2H_3O_2$ ( )
to 500 cm³ of 0.10 M $NaC_2H_3O_2$. What is the maximum amount of HCl that can be added to this solution without exceeding the capacity of the buffer?
   (a) 0.01 mol   (b) 0.05 mol
   (c) 0.10 mol   (d) 0.20 mol

21. If one were to prepare a buffer by using HSO_3^- and SO_3^{2-}, ( )
its pH would probably be in the range of ($K_a\ HSO_3^- = 3 \times 10^{-8}$)
   (a) 6.5–8.5   (b) 10–12
   (c) 2.0–4.0   (d) 3 ± 1

22. $CuCl_2$ is prepared from $Cu(OH)_2$ by adding 0.10 M HCl to ( )
0.20 mol of solid $Cu(OH)_2$. How much HCl should be added?
   (a) 500 cm³   (b) 1000 cm³
   (c) 2000 cm³   (d) some other volume

23. The most appropriate equation for the reaction referred to ( )
in Question 22 is
   (a) $Cu(OH)_2(s) + 2\ H^+(aq) \rightarrow Cu^{2+}(aq) + 2\ H_2O$
   (b) $Cu^{2+}(aq) + 2\ Cl^-(aq) \rightarrow CuCl_2(s)$
   (c) $H^+(aq) + OH^-(aq) \rightarrow H_2O$
   (d) $Cu(H_2O)_4^{2+}(aq) \rightarrow Cu(H_2O)_3(OH)^+(aq) + H^+(aq)$

20–Acid-Base Reactions

24. The metal sulfides CoS, CuS, and FeS have solubility ( )
products of 10^{-21}, 10^{-25}, and 10^{-17}, respectively, The most soluble in acid will be
    (a) CoS   (b) CuS
    (c) FeS   (d) none of these

25. Consider the three indicators methyl red ($K_a = 10^{-5}$), litmus ($K_a = 10^{-7}$), and phenolphthalein ($K_a = 10^{-9}$). Which indicator would you use to titrate
    (a) NH_3 with HCl?_____
    (b) NaOH with HCl?_____
    (c) $C_2H_3O_2^-$ with HNO_3?_____
    (d) $HC_2H_3O_2$ with NaOH?_____
    (e) HCN with NH_3?_____

26. $K_a = 1 \times 10^{-6}$ for a certain indicator HIn. The undis- ( )
sociated HIn has a red color; the anion, In^-, a yellow color. The color of the indicator in a solution of pH = 8.0 will be
    (a) red   (b) orange
    (c) yellow   (d) green

27. $SO_2(g)$ may be prepared by adding a strong acid to a ( )
concentrated solution of sodium sulfite and heating. The net reaction is

$$SO_3^{2-}(aq) + 2 H^+(aq) \rightarrow H_2O + SO_2(g)$$

The procedure works well because
    (a) H_2SO_3 is a weak acid
    (b) K for the reaction is large
    (c) the solubility of SO_2 in H_2O is lower at higher temperatures
    (d) all the above

28. A sample of $BaSO_4(s)$ is contaminated with a small amount ( )
of $BaCO_3(s)$. The best way to purify the $BaSO_4$ is to
    (a) wash it with HCl(aq) to dissolve away the $BaCO_3$
    (b) wash with water to dissolve the $BaCO_3$ and leave behind the insoluble $BaSO_4$ (K_{sp} for $BaCO_3 = 10^{-9}$)
    (c) wash with NaOH(aq) to dissolve away the $BaSO_4$
    (d) dissolve both solids in water and then add H_2SO_4 to precipitate the $BaSO_4(s)$

29. In liquid ammonia, self-ionization occurs: $2 NH_3 \rightleftarrows NH_4^+ +$ ( )
NH_2^-, $K = 10^{-30}$. In this solvent, an acid might be
    (a) NH_3
    (b) NH_4^+

(c) any species which will form NH_4^+
(d) any of these

Problems

30. To prepare a buffer, 50.0 cm³ of 0.120 M NaOH was added to 30.0 cm³ of 0.250 M acetic acid, $HC_2H_3O_2$ ($K_a = 1.8 \times 10^{-5}$).
    (a) Write the net ionic equation for the reaction.
    (b) How many moles of $C_2H_3O_2^-$ are present in the final solution?
    (c) How many moles of $HC_2H_3O_2$ are present?
    (d) What is the pH of this buffer?

31. The dissociation constants of the weak acids $HC_2H_3O_2$ and H_2CO_3 are 1.8×10^{-5} and 4.2×10^{-7}, respectively. Calculate K for each of the following reactions ($K_w = 1.0 \times 10^{-14}$).
    (a) $C_2H_3O_2^-(aq) + H^+(aq) \rightarrow HC_2H_3O_2(aq)$
    (b) $HC_2H_3O_2(aq) + OH^-(aq) \rightarrow C_2H_3O_2^-(aq) + H_2O$
    (c) $H_2CO_3(aq) + C_2H_3O_2^-(aq) \rightarrow HCO_3^-(aq) + HC_2H_3O_2(aq)$

32. Write net ionic equations to describe what happens when
    (a) a strong acid is added to a solution of Na_2CO_3.
    (b) a solution of ammonium sulfate is heated with NaOH.
    (c) hydrochloric acid is added to a solution of NH_3.
    (d) $ZnCO_3(s)$ dissolves in a solution of hydrochloric acid.

*33. As much as three-fourths of the nitric acid produced in the U.S. is converted, by reaction with ammonia, to ammonium nitrate. About 6.4×10^9 kg of nitric acid were produced in 1975.
    (a) Write equations for each of the reactions involved in the major commercial preparations of NH_3, HNO_3, and NH_4NO_3.
    (b) Estimate the amount of NH_4NO_3 produced in 1975.
    (c) Determine the major uses for NH_4NO_3.

*34. Demonstrate, by a suitable calculation, whether it is feasible to remove deposits of $Fe(OH)_3$ from a porcelain sink by washing with a dilute solution of acetic acid or of hydrochloric acid. What is the origin of these unsightly deposits?

SELF-TEST ANSWERS

1. T ($1/K_b < 1/K_w$.)
2. F (Solution contains $C_2H_3O_2^-$; hence, it is basic.)
3. T (Again, *complete conversion of reactants.*)
4. T (Net reaction is reverse of dissociation of $HC_2H_3O_2$.)

5. T (One mole H⁺ supplied by one mole of acid.)
6. F .(Less than 7; solution will contain the weak conjugate acid.)
7. T (Equivalent amounts of $HC_2H_3O_2$, $C_2H_3O_2^-$ are present.)
8. F (No capacity to absorb H^+.)
9. F (Strong base: the reaction is $NH_4^+ + OH^- \rightarrow NH_3(g) + H_2O$.)
10. T (CO_2 (an acid anhydride — see Chapter 17) gives acidic water solution; so here, $OH^- + H_2CO_3 \rightarrow H_2O + HCO_3^-$.)
11. c (The acid is weak — see Table 19.4 — so predominant species is HF, not H^+.)
12. c (NH_3 is a weak base.)
13. d (What would be the value of K?)
14. c (For the reverse reaction, you should recognize $1/K_b$ or K_a/K_w as the equilibrium constant.)
15. c (Don't forget the total volume.)
16. c (SO_4^{2-} is not very basic; K_b is very small.)
17. c
18. d (Depending on whether it reacted to HCO_3^- or CO_2.)
19. a ($K_b = [HX][OH^-]/[X^-]$, so $[OH^-] = 10^{-10}$, $[H^+] = 10^{-4}$.)
20. b (Equal to number of moles of weak base.)
21. a (For $[H^+] \approx K_a$.)
22. d (4000 cm³; note that one mole of solid gives two moles of OH^-.)
23. a (Followed perhaps by evaporation, answer b.)
24. c
25. (a) MR (b) any one (c) MR (d) P (e) None would work
26. c ($[In^-]/[HIn] = K_a/[H^+] = 10^2$.)
27. d
28. a (Both are insoluble in water: see solubility rules, Chapter 18.)
29. d (Using the Brönsted definition, Chapter 19, NH_3 and NH_4^+ are sources of protons.)
30. (a) $HC_2H_3O_2(aq) + OH^-(aq) \rightarrow C_2H_3O_2^-(aq) + H_2O$
    (b) $0.120 \frac{mol}{dm^3} \times 0.0500\ dm^3 = 6.00 \times 10^{-3}\ mol\ C_2H_3O_2^-$
    (c) $7.50 \times 10^{-3} - 6.00 \times 10^{-3} = 1.50 \times 10^{-3}\ mol\ HC_2H_3O_2$
    (d) $[H^+] = K_a \times \frac{[HC_2H_3O_2]}{[C_2H_3O_2^-]} = 1.8 \times 10^{-5} \frac{(1.50 \times 10^{-3})}{(6.00 \times 10^{-3})} =$
    $= 4.5 \times 10^{-6}$
    pH = 5.35
31. (a) $K = 1/1.8 \times 10^{-5} = 5.6 \times 10^4$
    (b) $K = 1/K_b = K_a/K_w = 1.8 \times 10^{-5}/1.0 \times 10^{-14} = 1.8 \times 10^9$
    (c) $K = K_a\ H_2CO_3/K_a\ HC_2H_3O_2 = 4.2 \times 10^{-7}/1.8 \times 10^{-5} =$
    $= 2.3 \times 10^{-2}$
32. (a) $CO_3^{2-}(aq) + 2\ H^+(aq) \rightarrow CO_2(g) + H_2O$
    (b) $NH_4^+(aq) + OH^-(aq) \rightarrow NH_3(g) + H_2O$

(c) $NH_3(aq) + H^+(aq) \rightarrow NH_4^+(aq)$
(d) $ZnCO_3(s) + 2 H^+(aq) \rightarrow Zn^{2+}(aq) + H_2O + CO_2(g)$

*33. (a) See the discussion, in the text, of the Haber and Ostwald processes.
(b) 6.0×10^9 kg, assuming 100% yield
(c) An encyclopedia of science and technology should answer the question.

*34. You would want to determine the magnitude of K for a reaction such as

$$Fe(OH)_3(s) + 3 HC_2H_3O_2(aq) \rightarrow Fe^{3+}(aq) + 3 H_2O + 3 C_2H_3O_2^-(aq)$$

If $K \geq 1$, removal should be thermodynamically feasible.

SELECTED READINGS

See the Readings listed for the preceding chapter.
For practice in acid-base stoichiometry, see the problem manuals listed in the Preface.

21
COMPLEX IONS; COORDINATION COMPOUNDS

QUESTIONS TO GUIDE YOUR STUDY

1. What is a *complex ion*? A *coordination compound*? What special properties do they possess?
2. Where are you likely to encounter complex ions? What elements are most likely to form complexes?
3. What kind of experimental support do we have for the existence of complexes? How could we show that they exist in the solid state? In solution?
4. What is the nature of the bonding in these species? What geometries do you associate with various complexes?
5. How does the bonding and geometry of complexes account for their properties?
6. How do you account for the formation of a complex ion in terms of ΔH and ΔS?
7. How can you decide on the relative stabilities of complexes? How can you measure the extent to which one species is formed at the expense of another? How is this quantitatively expressed?
8. How can you change the extent of a reaction in which a complex is formed or decomposed?
9. When a complex ion takes part in a reaction, how does the rate depend on the nature of the complex?
10. What uses have been found for complexes? Can you justify their study? What natural products contain complex ions?

11.

12.

YOU WILL NEED TO KNOW

Concepts

1. How to write electron configurations and draw orbital diagrams for transition metal atoms and ions — Chapters 6–8.
2. The geometries of atomic orbitals — Chapter 6; and of hybrid atomic orbitals; the overall shapes of tetrahedra and octahedra — Chapter 8.
3. How to write Lewis structures for molecules and ions — Chapter 8.
4. The general principles of equilibrium (e.g., effects of concentration changes; Rule of Multiple Equilibria) — Chapter 15 (and applied in Chapters 18–20).
5. The Lewis concept of acids and bases — Chapter 19.
6. Some of the methods of structure determination — see, e.g., Chapter 8 of this Guide.

Math

1. How to calculate K using the Rule of Multiple Equilibria — Chapter 15 (and illustrated in 20); how to use K.

CHAPTER SUMMARY

This chapter offers the opportunity for you to consider the real nature of most species in water solution. In particular, transition metal ions are generally complexed by the solvent or by some other, more strongly basic, ligand (basic in the Lewis sense). You have already encountered complexes; recall that the explanation for the acidity of metal ions (Chapter 19) involved aquo complexes. And you may have observed directly how their formation will permit the separation of metal ions in qualitative analysis. Were these reasons not enough justification for studying complexes, we could also point out the existence of many biologically important compounds in which central atoms are bonded to more than their expected share of other atoms. The energy-storing cytochromes are protein-encapsulated complexes of iron, with the metal ion bonded to six other atoms. Many common drugs, aspirin among them, may derive their potency from their ability to serve as chelating agents.

The very existence of complexes, particularly those of the transition metals, depends on there being low-energy, closely spaced orbitals available for donor pairs of electrons to go into. The d orbitals and the s and p orbitals of adjacent principal energy levels are thus used most often. For most of the transition metals, the d orbitals are at least partially empty and therefore available for bonding.

Crystal field theory is rather successful in predicting how the relative energies of these orbitals on the central atom change with the nature of the ligands. These orbital energy changes in turn provide an explanation for the dependence of properties such as color, behavior in a magnetic field, relative stability (measured, for example, by K_d) . . . on the nature of the ligands.

A relatively simple and generally successful picture of the bonding in complexes is given by valence bond theory. For the complexes we have considered in this chapter, the following correlations can be made: When two bonds are formed by the central atom, that atom employs two equivalent, hybrid atomic orbitals of the type sp (such hybrids are directed along a straight line away from the central atom). When four bonds form, hybrids are either sp^3 (with tetrahedral geometry) or dsp^2 (square), depending on what orbitals are available. When six bonds are formed, six equivalent hybrids are derived from two d, one s, and three p orbitals (d^2sp^3). There are other hybrids that could be described; there are other coordination numbers (numbers of atoms bonded to the central one). The four mentioned are the most frequently encountered. Note that electron pair repulsion ideas would allow you to predict most of the geometries equally as well, knowing simply the coordination number (see Chapter 8).

BASIC SKILLS

1. Deduce the structure of a complex ion, given the properties of a compound formed by that ion.

This skill is illustrated by Example 21.1 and, most simply, by Problems 21.1 and 21.8. Problems 21.6, 21.7, 21.25, and 21.26 involve the same principle. In working these problems, remember that:

     a. square brackets have been used to set off the formula of the complex ion.

     b. Ag^+ ion at $0°C$ forms a precipitate only with halide ions outside the complex.

     c. the conductivities of coordination compounds are similar to those of simple salts of the same valence type (Table 21.1).

286 • 21–Complex Ions; Coordination Compounds

2. **Given the composition of a complex, sketch its geometry, including any geometrical isomers.**

In essence, this skill involves three steps.

a. *Determine the coordination number.* This is ordinarily easy; the species $Co(NH_3)_6^{3+}$, $Zn(NH_3)_4^{2+}$, and $Ag(NH_3)_2^+$ obviously have coordination numbers of 6, 4, and 2 respectively. If one or more of the ligands is a chelating agent, the coordination number may not be so obvious. Consider, for example, the complexes $Co(NH_3)_4(en)^{3+}$, $Co(NH_3)_2(en)_2^{3+}$, and $Co(en)_3^{3+}$. Since each ethylenediamine molecule bonds to Co^{3+} at two different places, the coordination number in each of these species is 6.

b. *Relate the coordination number to the geometry (apply electron pair repulsion ideas of Chapter 8).* If the coordination number is 2, the complex is linear; a coordination number of 6 corresponds to an octahedral complex. For coordination number 4, two geometries turn out to be possible: tetrahedral or square. There is no simple way to predict which geometry will prevail with a particular ion.

c. *Having decided upon the geometry, sketch the complex.* This is entirely straightforward for linear and tetrahedral complexes. With square complexes you must keep in mind the possibility of geometrical isomerism in complexes of the form MA_2B_2, where A and B are different ligands. Writing all of the isomers for an octahedral complex can be a little tougher. Example 21.2 suggests a logical approach to follow. Note in particular that for any given site in an octahedral complex, there is one site that is *trans* to it and four that are *cis*. This remains true regardless of which site you choose as your starting point.

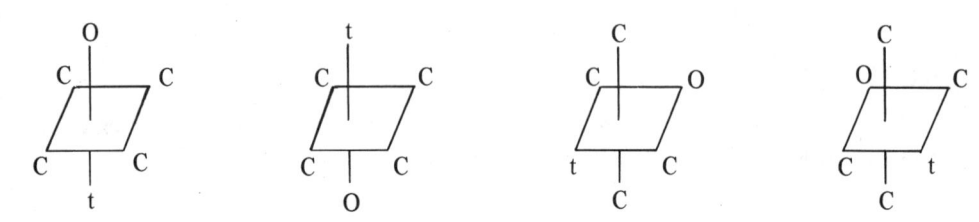

o = original site; c = *cis* position; t = *trans* position

Problems 21.2, 21.11, 21.12, 21.30, and 21.31 illustrate these principles; Problems 21.13 and 21.32 are a little more subtle but require the same skill.

3. **Given the composition of a complex ion and its geometry, draw an orbital diagram for the electrons around the central metal atom (valence bond model).**

Several such diagrams are given in the text; the process used to obtain them is described there and illustrated in Example 21.3. Note that:

a. the orbital diagram ordinarily includes only those electrons beyond the nearest noble gas core of electrons. In the structures on these pages, the 18 argon electrons are not shown.
b. the number of orbitals filled by the bonding electrons, contributed by the ligands, is always equal to the coordination number, regardless of the nature of the ligands.

A systematic approach to drawing orbital diagrams might involve the four steps listed below. For simplicity, we assume here that the metal involved is in the first transition series (atomic number 21–30), but the same general approach can be extended to other metals. (Obtaining electron structures by analogy was illustrated in Chapter 7.)

    a. Decide which orbitals are filled by the bonding electrons. This requires only that you know the geometry of the complex.

Coord. no.	Geometry	Hybridization	Orbitals filled (first transition series)
2	linear	sp	one 4s, one 4p
4	square	dsp^2	one 3d, one 4s, two 4p
4	tetrahedral	sp^3	one 4s, three 4p
6	octahedral	d^2sp^3	two 3d, one 4s, three 4p

    b. Fill these orbitals with the bonding electrons and leave them there!
    c. Decide how many electrons are contributed to the orbital diagram by the central metal ion. To do this, start with the atomic number of the metal, subtract those electrons lost in forming the ion, and finally take away the 18 Ar electrons.
    d. If possible, fit the electrons left after (c) into the 3d orbitals, following Hund's rule. Occasionally, you may find that there are too many electrons to fit into the available 3d orbitals. If this happens, put the overflow into the next highest empty orbital.

The application of this approach is shown in the following example.

Draw orbital diagrams for the square and tetrahedral complexes of Ni^{2+}.

We start by noting that in the square complex the hybridization is dsp^2, while in the tetrahedral complex it is sp^3 (step a). These orbitals are then filled by the bonding electrons (step b).

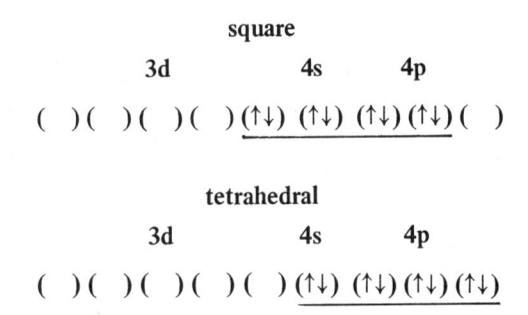

square

3d   4s   4p

( )( )( )( ) (↑↓) (↑↓) (↑↓)(↑↓) ( )

tetrahedral

3d   4s   4p

( )( )( )( )( ) (↑↓) (↑↓) (↑↓) (↑↓)

Since the atomic number of Ni is 28, there are 28 electrons in the Ni atom. In the Ni^{2+} ion, there are $28 - 2 = 26$ e^-. Subtracting the 18 argon electrons gives us $26 - 18 = 8$ e^- contributed to the orbital diagram by the Ni^{2+} ion (step c). These are just enough to fill all the available 3d orbitals in the square complex. In the tetrahedral complex, we distribute the 8 electrons among the five 3d orbitals so as to give the maximum number of unpaired electrons (step d).

square Ni^{2+}

3d   4s   4p

(↑↓) (↑↓) (↑↓) (↑↓)(↑↓) (↑↓) (↑↓)(↑↓) ( )

tetrahedral Ni^{2+}

3d   4s   4p

(↑↓) (↑↓) (↑↓) (↑)(↑) (↑↓) (↑↓) (↑↓) (↑↓)

Problems 21.3, 21.14, 21.15, 21.33, and 21.34 require the use of this skill.

4. For any octahedral complex, draw diagrams for the "high spin" and "low spin" forms (crystal field model).

The process involved here is indicated in Example 21.4. See Problems 21.4, 21.16, and especially 21.35. Note that two steps are involved:

    a. The five 3d orbitals are split into a lower energy group of three orbitals and a higher energy group of two orbitals.
        b. The d electrons in the metal ion are located:
            — according to Hund's rule for the "high spin" complex, ignoring the energy difference between the two groups of orbitals.

Basic Skills • 289

– preferentially in the lower energy orbitals for the "low spin" complex.

If you work Problem 21.35, you should find that with one, two, three, eight, nine, or ten d electrons, these two distributions lead to the same structure. Only if there are four to seven electrons to distribute can we distinguish high and low spin forms.

5. Given the dissociation constant for a complex ion:

   a. relate the concentration of "free" metal ion, complex ion, and ligand.

A typical calculation is given in Example 21.5. See Problems 21.5, 21.19, and 21.38. No new principle is involved here. The equilibrium constant for the dissociation of a complex ion is handled in exactly the same way as that for the dissociation of a weak acid (Chapter 19).

   b. Using the Rule of Multiple Equilibria, calculate K for the process of dissolving an insoluble compound in a complexing agent and the solubility of the compound in the complexing agent.

See Example 21.6 and Problems 21.20, 21.21, 21.39, and 21.40. Note that ordinarily:
– $K = K_{sp}/K_d$, where K_{sp} is a solubility product constant and K_d is the dissociation constant of the complex ion.
– The "solubility" is equal to the concentration of complex ion in a solution of known concentration of complexing agent.

6. Write net ionic equations for reactions involving the formation of complex ions.

Complex ions may be formed by:

   a. *direct reaction between metal ion and ligand in solution.* An example would be the reaction that occurs when aqueous ammonia is added to a solution of a Cu^{2+} salt, such as $CuSO_4$. This is described by the equation:

$$Cu^{2+}(aq) + 4\ NH_3(aq) \rightarrow Cu(NH_3)_4^{2+}(aq)$$

   b. *reaction between the ligand in solution and an insoluble solid containing the metal ion.* An example is the reaction that occurs when insoluble $Cu(OH)_2$ is treated with aqueous ammonia. The Cu^{2+} ion is brought into solution as the $Cu(NH_3)_4^{2+}$ complex:

290 • 21—Complex Ions; Coordination Compounds

$$Cu(OH)_2 (s) + 4\ NH_3 (aq) \rightarrow Cu(NH_3)_4^{2+}(aq) + 2\ OH^-(aq)$$

Problems. 21.22 and 21.41 illustrate this skill in simplest form. Parts of other problems require that you write net ionic equations to explain what is going on.

SELF-TEST

True or False

1. Of the two elements, calcium and copper, the more likely to ( ) form coordination compounds is copper.

2. Of the two ions, CN^- and SO_4^{2-}, the more likely to be ( ) found in metal complexes is the sulfate ion, the weaker base.

3. The sign of the entropy change for the reaction ( )

$$Ni^{2+}(aq) + dimethylglyoxime \rightarrow chelate(s)$$

is expected to be negative.

4. One would predict that $[Pt(NH_3)_4]Cl_2$ is more soluble in ( ) water than is $[Pt(NH_3)_2 Cl_2]$.

5. The smaller the value of the dissociation constant for a ( ) complex, the weaker the coordinate bonds in the complex.

6. The strength of the coordinate covalent bond is comparable ( ) to that of the intermolecular forces (dispersion, dipole . . .).

7. Any complex ion with a coordination number of four for ( ) the central atom has a tetrahedral structure.

8. Geometric (*cis, trans*) isomerism is not observed in tetra- ( ) hedral complexes.

9. For a complex ion, *labile* means exactly the opposite of ( ) stable.

10. The color often associated with complexes is best explained ( ) by the approach of valence bond theory.

Multiple Choice

11. In the $NiCl_4^{2-}$ ion, the total number of electrons around the ( ) Ni, including bonding pairs, is

(a) 26 (b) 28
(c) 34 (d) 36

12. The formula $PdCl_2(OH)_2^{2-}$ is known to represent two ( )
different ions. The hybrid orbitals occupied by the bonding electrons
are
   (a) sp (b) sp^3
   (c) dsp^2 (d) d^2sp^3

13. The hybridization of gold in $Au(NH_3)_2^+$ is ( )
   (a) d^2sp^3 (b) dsp^2
   (c) sp^3 (d) sp

14. The maximum number of possible geometric isomers for a ( )
complex having sp^3 hybridization would be
   (a) two (b) three
   (c) four (d) none

15. Geometric isomers would be expected for ( )
   (a) $Zn(NH_3)_4^{2+}$ (b) $Zn(H_2O)_2(OH)_2$
   (c) $Co(NH_3)_3Cl_2Br$ (d) $Au(NH_3)_2^+$

16. Complex ions are held together by ( )
   (a) marital bonds (b) municipal bonds
   (c) coordinate covalent bonds (d) James bonds

17. Complex ions of coordination number six have a geometric ( )
structure that is
   (a) linear (b) square
   (c) tetrahedral (d) octahedral

18. It is known that the sulfhydryl group, —SH, forms strong ( )
coordinate bonds to certain heavy metal ions. Which of the following
do you expect to be the best chelating agent for heavy metal ions?
   (a) CH_3-SH (b) $H-SH$
   (c) $CH_3-S-S-CH_3$ (d) $HS-CH_2-\underset{\underset{SH}{|}}{CH}-CH_2-OH$

19. Which species in the following reaction acts as a Lewis acid? ( )

$CuSO_4(s) + 4 NH_3(aq) \rightarrow Cu(NH_3)_4^{2+}(aq) + SO_4^{2-}(aq)$

   (a) Cu^{2+} (b) NH_3
   (c) SO_4^{2-} (d) $Cu(NH_3)_4^{2+}$

20. AgCl may be brought into water solution by the addition of ( )
   (a) NaCl (b) $AgNO_3$
   (c) H_2O (d) NH_3

292 • 21–Complex Ions; Coordination Compounds

21. Of the two complexes, $Cu(H_2O)_4^{2+}$ and $Cu(NH_3)_4^{2+}$, the ( )
second is the more stable. This means that
   (a) $Cu(NH_3)_4^{2+}$ would be the stronger acid
   (b) ethylenediamine would replace H_2O faster than it would replace NH_3
   (c) $Cu(NH_3)_4^{2+}$ has a smaller dissociation constant
   (d) $Cu(NH_3)_4^{2+}$ has a larger dissociation constant

22. The electronic structure for the central atom in $Co(en)_2Cl_2^+$ ( ) is
   (a) (↑↓)(↑↓)(↑↓)( ) |(↑↓) (↑↓) (↑↓)(↑↓)( )
   (b) (↑↓)(↑↓)(↑↓) |(↑↓) (↑↓) (↑↓) (↑↓)(↑↓)(↑↓)
   (c) (↑↓)(↑↓)(↑↓)(↑↓)(↑) |(↑↓) (↑↓)(↑↓)(↑↓)
   (d) (↑↓)(↑)(↑)(↑)(↑)|(↑↓) (↑↓)(↑↓)(↑↓)

23. A compound has the empirical formula $CoCl_3 \cdot 4\,NH_3$. One ( )
mole of it yields one mole of AgCl on treatment with excess $AgNO_3$.
Ammonia is not removed by treatment with concentrated sulfuric
acid. The formula of the compound is best represented by
   (a) $Co(NH_3)_4Cl_3$ (b) $[Co(NH_3)_4]Cl_3$
   (c) $[Co(NH_3)_3Cl_3]NH_3$ (d) $[Co(NH_3)_4Cl_2]Cl$

24. The dissociation constant for the complex ion $Zn(NH_3)_4^{2+}$ ( )
is the equilibrium constant for the reaction represented by the
equation
   (a) $Zn^{2+}(aq) + 4\,NH_3(aq) \rightleftarrows Zn(NH_3)_4^{2+}(aq)$
   (b) $Zn(NH_3)_4^{2+}(aq) + H_2O \rightleftarrows Zn(NH_3)_3(H_2O)^{2+}(aq) + NH_3(aq)$
   (c) $Zn(NH_3)_4^{2+} + 2\,e^- \rightleftarrows Zn(s) + 4\,NH_3(aq)$
   (d) $Zn(NH_3)_4^{2+}(aq) + 4\,H_2O \rightleftarrows Zn(H_2O)_4^{2+}(aq) + 4\,NH_3(aq)$

25. Of the following 1.0 M solutions, which has the greatest ( )
molar entropy?
   (a) NaCl   (b) $CuCl_2$
   (c) $AlCl_3$ (d) $[Co(NH_3)_5Cl]Cl_2$

26. To account for the fact that $Fe(NO_3)_3$ dissolves in water to ( )
give an acidic solution, we might write
   (a) $H_2O \rightleftarrows H^+(aq) + OH^-(aq)$
   (b) $Fe(NO_3)_3(s) \rightleftarrows Fe^{3+}(aq) + 3\,NO_3^-(aq)$
   (c) $Fe^{3+}(aq) + 3\,H_2O \rightleftarrows Fe(OH)_3(s) + 3\,H^+(aq)$
   (d) $Fe(H_2O)_6^{3+}(aq) \rightleftarrows Fe(H_2O)_5(OH)^{2+}(aq) + H^+(aq)$

27. How might you show experimentally that a particular ( )
complex is square?

(a) two isomers might be isolated; they would show different lability
(b) x-ray diffraction for the complex in the solid state would reveal the geometry
(c) magnetic measurements might distinguish between dsp^2 and sp^3 orbitals
(d) all of the above

28. When $[Ni(NH_3)_4]^{2+}$ is treated with concentrated HCl, two ( ) compounds having the same formula, $Ni(NH_3)_2Cl_2$, designated I and II are formed. Compound I can be converted to compound II by boiling in dilute HCl. A solution of I reacts with oxalic acid, $H_2C_2O_4$, to form $Ni(NH_3)_2(C_2O_4)$. Compound II does not react with oxalic acid. Compound II is
    (a) the same as compound I   (b) the *cis* isomer
    (c) the *trans* isomer   (d) tetrahedral in shape

29. The central atom in $[Co(NH_3)_5Cl](NO_3)_2$ can be assigned ( ) a charge of
    (a) +1   (b) +2
    (c) +3   (d) something else

30. The hybrid orbitals used by $_{13}Al$ in the complex $AlCl_4^-$ are ( ) expected to be
    (a) sp   (b) sp^3
    (c) dsp^2 (d) no hybrids used

Problems

31. Two complex compounds have the same formula, $Rh(NH_3)_4Br_2Cl$. One is yellow, the other orange. Both compounds react with $AgNO_3$ solution to precipitate exactly one mole of AgCl for every mole of complex; no AgBr precipitates.
    (a) What is(are) the formula(s) of the complex ion(s) present?
    (b) Draw structures for the two complex ions.

32. A certain coordination compound has the formula $[Co(en)_2Cl_2]Cl$. Answer the following questions about the cation in this compound:
    (a) What is the coordination number of the central cobalt atom?
    (b) Sketch the two isomers of the complex cations, labelling as *cis* and *trans*.
    (c) Draw the orbital diagram for Co^{3+} in the complex (at. no. Co = 27).

33. The following four complexes of Cr^{3+} have been isolated. All have three unpaired d electrons (at. no. Cr = 24).

	(a)	(b)	(c)	(d)
Formula	$Cr(NH_3)_3Cl_3$	$Cr(NH_3)_4Cl_3$	$Cr(NH_3)_4Cl_3$	$Cr(NH_3)_6Cl_3$
Color	green	brown	violet	yellow
Electrical conductivity	like CH_3OH	like NaCl	like NaCl	like $Al(NO_3)_3$

Sketch a structure for each complex.

*34. The presence of iron salts in a water supply may lead to the deposit of reddish-brown $Fe(OH)_3$ on a porcelain sink. The removal of such unattractive deposits is commonly performed by washing with a solution of oxalic acid, $H_2C_2O_4$. (Caution!)

Using the data below and from the text, show by calculation which of the two equations is the more likely to represent the mechanism of dissolving the $Fe(OH)_3$.

1) acid-base mechanism:

$$2\ Fe(OH)_3(s) + 3\ H_2C_2O_4(aq) \rightarrow 2\ Fe^{3+}(aq) + 6\ H_2O + 3\ C_2O_4^{2-}(aq)$$

2) complex-formation mechanism:

$$2\ Fe(OH)_3(s) + 6\ H_2C_2O_4(aq) \rightarrow 2\ Fe(C_2O_4)_3^{3-}(aq) + 6\ H_2O + 6\ H^+(aq)$$

Data:

$Fe(C_2O_4)_3^{3-}$ $K_d = 1 \times 10^{-20}$

$H_2C_2O_4$ $K_1 = 6 \times 10^{-2}$; $K_2 = 6 \times 10^{-5}$
(first and second K_a's)

SELF-TEST ANSWERS

1. **T** (Transition metal.)
2. **F** (The base donates the bonding electrons.)
3. **F** (Consider that several moles of water are produced. The water is in the aquo complex, usually indicated merely by (aq). $\Delta S > 0$ tends to drive the reaction, i.e., make $\Delta G < 0$; see Chapter 14.)
4. **T** (The first gives ions in solution; the second is a nonelectrolyte.)
5. **F** (Stronger. How does K depend on ΔH? See Chapter 15.)
6. **F** (Many complexes are very stable!)
7. **F** (May be square.)

8. T (So here is one way of distinguishing experimentally between square and tetrahedral complexes.)
9. F (Labile refers to rate; stable, to equilibrium position.)
10. F (Crystal field or ligand field theory.)
11. c (For Ni^{2+} and four pairs of electrons from the four Cl^- ions.)
12. c (*Cis-trans* isomers.)
13. d
14. d (See Question 8 above. All positions are equidistant from one another.)
15. c (Complex b is tetrahedral.)
16. c
17. d (With six vertices.)
18. d (This would permit ring formation. How would you treat heavy metal poisoning?)
19. a (Electron pair acceptor — Chapter 19.)
20. d (To form a complex. What calculation would demonstrate its feasibility?)
21. c
22. b (en is a chelating agent with two pairs of electrons to donate.)
23. d
24. d (Choice b would be only the first step in the overall dissociation that is usually represented by K. Note that you generally omit the water from the equation in d.)
25. c (Giving 4 mol of ions, compared with 3 for the complex.)
26. d (No precipitate forms, as c would indicate. See Chapter 19.)
27. d
28. c (Forming the chelate generally requires easy access to *cis* positions.)
29. c (The complex itself carries a charge of +2; expecting chlorine to be −1 leaves Co^{3+}. Common ions, such as NO_3^- and Cl^-, have been seen repeatedly.)
30. b (Al^{3+} has available 3s, 3p, etc. Fill lowest first.)
31. (a) $Rh(NH_3)_4Br_2^+$

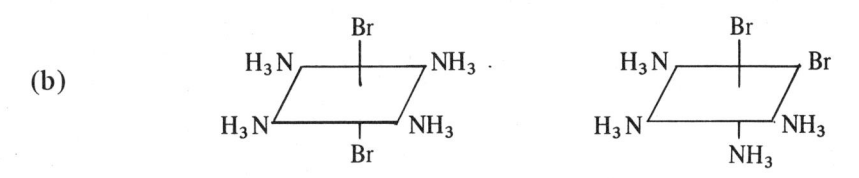

(b)

$$\begin{array}{c} Br \\ H_3N \diagdown | \diagup NH_3 \\ H_3N \diagup | \diagdown NH_3 \\ Br \end{array} \qquad \begin{array}{c} Br \\ H_3N \diagdown | \diagup Br \\ H_3N \diagup | \diagdown NH_3 \\ NH_3 \end{array}$$

32. (a) 6

(b)

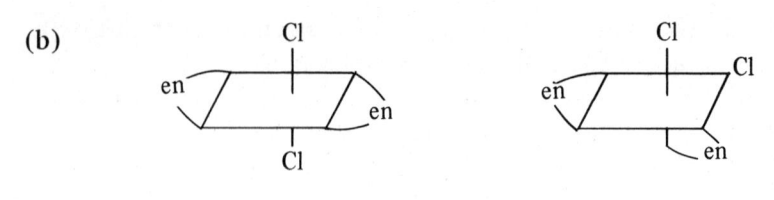

(en must attach at cis positions)

trans   cis

(c)   3d   4s   4p

[Ar] (↑↓) (↑↓) (↑↓) (↑↓) (↑↓) (↑↓) (↑↓) (↑↓) (↑↓)

33. (a)   (b)

(c)   (d)

*34. *Carefully* apply the Rule of Multiple Equilibria. Recall Problem 34 in the Self-Test of Chapter 20.

Would you believe, for the complex formation, $K = \dfrac{(K_{sp})^2 (K_1)^6 (K_2)^6}{(K_d)^2 (K_w)^6} \approx 10^{17}$?

SELECTED READINGS

Alternative discussions, including more advanced treatment of valence bond and ligand field theory:

Basolo, F., *Coordination Chemistry*, New York, W. A. Benjamin, 1964.

Cotton, F. A., Ligand Field Theory, *Journal of Chemical Education* (September 1964), pp. 466-476.

Pauling, L., *The Nature of the Chemical Bond*, Ithaca, N. Y., Cornell University Press, 1960.

From reaction mechanisms to biological and medicinal complexes:

Bailar, J. C., Jr., Some Coordination Compounds in Biochemistry, *American Scientist* (September-October 1971), pp. 586-592.

House, J. E., Jr., Substitution Reactions in Metal Complexes, *Chemistry* (June 1970), pp. 11-14.

Jones, M. M., Therapeutic Chelating Agents, *Journal of Chemical Education* (June 1976), pp. 342-347.

Perutz, M. F., The Hemoglobin Molecule, *Scientific American* (November 1964), pp. 64-76.

Schubert, J., Chelation in Medicine, *Scientific American* (May 1968), pp. 40-50.

22

OXIDATION AND REDUCTION; ELECTROCHEMICAL CELLS

QUESTIONS TO GUIDE YOUR STUDY

1. What reactions have you encountered that can be classified as oxidation or reduction? What, for example, is a reducing flame? An oxidizing atmosphere?

2. How do you recognize what is oxidized and what is reduced in the equation for a redox reaction?

3. How do you write and interpret redox equations?

4. What occurs in an electrolytic cell? In a voltaic cell? (If you could watch the individual atoms, ions and molecules, what would you expect to see?)

5. How do fuel cells differ from other voltaic cells?

6. What energy effects are associated with reactions in electrochemical cells?

7. What generalizations can be made about what reactions may occur in an electrochemical cell? How do you predict what is oxidized, what is reduced? Are there any correlations to be made with the Periodic Table?

8. How is electrical energy quantitatively related to the masses of reacting species in electrochemical cells?

9. What can be said about the rates of redox reactions?

300 • 22–Oxidation and Reduction; Electrochemical Cells

10. How big, in terms of everyday experience, are the common electrical units: coulomb, ampere, volt, watt? (For example, how large a charge flows through your desk lamp per hour? Your electronic calculator?)

11.

12.

YOU WILL NEED TO KNOW

Concepts

1. How to draw Lewis structures — Chapter 8.
2. How to write and interpret balanced net ionic equations — Chapter 18.
3. How to interpret free energy changes and the sign of ΔG — Chapter 14.

Math

1. How to work problems in stoichiometry — Chapter 3.

CHAPTER SUMMARY

The last type of reaction that we discuss in this text is oxidation-reduction, which involves a transfer of electrons from a reducing agent such as Zn, H_2, or CH_4 to an oxidizing agent such as Zn^{2+}, H^+, or CO_2. The species which is oxidized increases in oxidation number (Zn → Zn^{2+}; O.N., 0 → +2); the species which is reduced decreases in oxidation number (H^+ → H_2; O.N., +1 → 0). Note that electron transfer may be complete, as in the reaction $Zn(s) + 2 H^+(aq) \rightarrow Zn^{2+}(aq) + H_2(g)$, where Zn atoms become ions. Or the transfer may be only partial: $CH_4(g) + 2 O_2(g) \rightarrow CO_2(g) + 2 H_2O(l)$, where carbon has only partly lost its bonding electrons to the more electronegative oxygen ($CH_4 \rightarrow CO_2$, O.N. of carbon, -4 → +4).

Oxidation numbers are assigned using a set of arbitrary rules. Perhaps the most important of these rules tells us that the sum of the oxidation numbers of all the atoms in a species is equal to the charge of that species. This rule can be applied to find oxidation numbers of elements in unfamiliar species. For example, for HBrO and BrO_3^- we find that the oxidation numbers of Br are:

HBrO: +1 + O.N. Br + (−2) = 0; O.N. Br = +1

BrO$_3^-$: O.N. Br + 3(−2) = −1; O.N. Br = +5

One method of balancing oxidation-reduction (redox) equations, outlined in the text, considers any given redox equation as the sum of two other "equations" — an oxidation and a reduction. (Note that the first step required is your recognizing that species undergoing oxidation and that undergoing reduction. Perhaps this is achieved most simply by the assignment of oxidation numbers to all atoms in all reactant and product species.) This particular approach will become very useful in Chapter 23. Once an equation is balanced, it can be used to perform stoichiometric calculations.

Many familiar redox reactions occur in the world around us. The rusting of iron, the combustion of fuels, and most of the processes involved in human metabolism are redox reactions. All of these reactions and many others that we carry out in the general chemistry laboratory (e.g., the reaction of metals with acids) involve a spontaneous electron transfer from reducing agent to oxidizing agent. Any such reaction can be adapted, at least in principle, to produce electrical energy in a voltaic cell. Most of the cells in current use (dry cell, storage battery) consume chemicals that are too expensive to make them practical, large-scale sources of electrical energy. Fuel cells, in which the chemical energy available from the combustion of fuels is converted directly to electrical energy, would seem to offer a partial solution to the energy crisis. The development of such cells has sufficiently progressed so that large-scale power plants are now planned.

Nonspontaneous redox reactions ($\Delta G > 0$) can be carried out by supplying electrical energy in an electrolytic cell. Many important metals (Na, Mg, Al) and industrial chemicals (Cl$_2$, NaOH) are produced in this way. To calculate the "yield" of products in an electrolytic cell from the quantity of electricity supplied, we use Faraday's laws. As we might expect, electrochemical processes seldom give a 100% yield of the desired product; side reactions divert an appreciable fraction of the electrons passing through the cell.

BASIC SKILLS

1. **Given the formula of an ion or molecule, determine the oxidation number of each atom.**

The rules for assigning oxidation number are given in the text and applied in Example 22.1. See also Problems 22.1, 22.5, and 22.25. Problems 22.6, 22.7, 22.26, and 22.27 are somewhat more subtle but involve the same principle.

22–Oxidation and Reduction; Electrochemical Cells

You should keep in mind that oxidation numbers are assigned in a quite arbitrary manner and do not have any direct physical meaning. Thus, you should not be disturbed if you find that an atom in a molecule or polyatomic ion has an oxidation number of 0 or $-1/2$.

2. **Given the formulas of products and reactants, balance a redox equation by the half-equation method.**

The method is described in the text in considerable detail. Perhaps another example would be helpful.

Balance the equation for the reaction

$$Sn^{2+}(aq) + NO_3^-(aq) \rightarrow Sn^{4+}(aq) + NO_2(g),$$

first in acidic and then in basic solution. _____ ; _____
The oxidation half-equation is:

$$Sn^{2+}(aq) \rightarrow Sn^{4+}(aq) + 2e^-$$

where the two electrons are required to balance the charges. The reduction half-equation involves the NO_3^- ion:

$$NO_3^-(aq) + 2H^+(aq) + e^- \rightarrow NO_2(g) + H_2O$$

Here, it was necessary to add one H_2O molecule to balance oxygen. This in turn required the addition of $2 H^+$ to the left to balance hydrogen. Finally, one electron was added to the left to balance charges.

To obtain the overall equation in acidic solution, we multiply the second half-equation by 2 and add it to the first. This has the effect of "canceling out" the electrons:

$$Sn^{2+}(aq) \rightarrow Sn^{4+}(aq) + 2\,e^-$$

$$\underline{2\,NO_3^-(aq) + 4\,H^+(aq) + 2\,e^- \rightarrow 2\,NO_2(g) + 2\,H_2O}$$

$$Sn^{2+}(aq) + 4\,H^+(aq) + 2\,NO_3^-(aq) \rightarrow Sn^{4+}(aq) + 2\,NO_2(g) + 2\,H_2O$$

To balance the equation in basic solution, we add $4\,OH^-$ ions to both sides:

$$Sn^{2+}(aq) + 2\,NO_3^-(aq) + 4\,H_2O \rightarrow Sn^{4+}(aq) + 2\,NO_2(g) + 2\,H_2O + 4\,OH^-(aq)$$

which simplifies to:

$$Sn^{2+}(aq) + 2\,NO_3^-(aq) + 2\,H_2O \rightarrow Sn^{4+}(aq) + 2\,NO_2(g) + 4\,OH^-(aq)$$

Refer to Problems 22.2, 22.10–22.14, and 22.30–22.34. A couple of hints:

a. O_2 in water solution is ordinarily reduced either to H_2O (acidic solution) or OH^- (basic solution).

b. It is quite possible for the same species to be both oxidized and reduced (Problem 22.33).

3. Relate the amount of electricity passed through an electrolytic cell to the amounts of substances produced or consumed at the electrodes.

Example 22.2 illustrates typical calculations of this kind. Note that you must know the half-equation for the electrode process or at least the change in oxidation number that is involved. In Example 22.3, the concept of gram equivalent mass in redox reactions is introduced. Note that, in any electrolysis:

$$\text{no. of moles } e^- = \text{no. of GEM}$$

Of the problems of this type at the end of the chapter, Problems 22.3, 22.19, 22.22, 22.39, and 22.42 are entirely analogous to the examples. Problems 22.20, 22.21, 22.40, and 22.41 involve the same principle but require additional conversions to arrive at a final answer.

4. Sketch electrolytic and voltaic cells corresponding to a given redox reaction, label anode and cathode, and trace the flow of current through the cell.

A simple electrolytic cell is shown in Figure 22.2. The $Zn-Cu^{2+}$ salt bridge cell shown in Figure 22.6 is typical of voltaic cells. Note that in both types of cells:

a. oxidation occurs at the anode, reduction at the cathode;
b. within the cell, anions move to the anode, cations to the cathode;
c. outside the cell, electrons move out of the anode and into the cathode.

See Problems 22.4, 22.16, 22.23, 22.36, and 22.43. Note that in designing voltaic cells it is frequently necessary to use an inert electrode; platinum is a safe (though expensive!) choice. Problems 22.24 and 22.44 are particularly instructive since they require you to connect a voltaic cell to an electrolytic cell.

SELF-TEST

True or False

1. In a redox reaction, the oxidizing agent gains electrons. ( )
2. The lowest oxidation state of a 5A element is −3. ( )
3. Metals frequently show negative oxidation numbers. ( )
4. A redox equation balanced in acidic solution can be ( ) converted to apply in basic solution by adding the proper number of H_2O molecules to both sides.
5. The purpose of the iron screen in the Downs cell is to ( ) prevent Na^+ and Cl^- ions from coming in contact with each other.
6. Cryolite may be added in the electrolytic process for ( ) making aluminum so that the electrolysis can be carried out at a lower temperature.
7. In the electrolysis of a water solution of NaCl, one mole of ( ) OH^- is produced for every mole of Cl^- consumed.
8. Complexing agents such as CN^- are used in many electro- ( ) plating processes to increase the concentration of metal ions.
9. In the electrolysis of a solution of $Ag(S_2O_3)_2^{3-}$, three ( ) moles of electrons are required to form one mole of silver.
10. In any cell, electrolytic or voltaic, the cathode is the ( ) negative electrode.

Multiple Choice

11. The oxidation number of Mn in the MnO_4^- ion is ( )
    (a) −2   (b) +6
    (c) +7   (d) +8

12. The oxidation number of P in H_3PO_4 is ( )
    (a) −3   (b) +1
    (c) +3   (d) +5

13. Using Figure 22.1, decide which one of the following is not ( ) a reasonable formula for an oxide of chromium:
    (a) Cr_2O (b) CrO
    (c) Cr_2O_3 (d) CrO_3

14. When the half-equation $I_2 + e^- \rightarrow I^-$ is balanced, the ( )
coefficients of I_2, e^-, and I^- are, respectively,
  (a) 1, 1, 1   (b) 1, 1, 2
  (c) 1, 2, 2   (d) 1, 0, 2

15. When the half-equation $HSO_3^- + H_2O \rightarrow SO_4^{2-} + H^+ + e^-$ is ( )
balanced, the coefficients, reading from left to right, are
  (a) 1, 1, 1, 1, 1   (b) 1, 1, 1, 2, 2
  (c) 1, 2, 1, 5, 2   (d) 1, 1, 1, 3, 2

16. When the half-equations in Questions 14 and 15 are ( )
combined, it is necessary to multiply the reduction half-equation by
_____ and the oxidation half-equation by _____ before adding.
  (a) 1, 1   (b) 1, 2
  (c) 2, 1   (d) 1, 3

17. In the electrolysis of molten magnesium chloride, the most ( )
appropriate equation for the anode reaction would be
  (a) $Mg^{2+} + 2 e^- \rightarrow Mg(s)$
  (b) $2 H_2O + 2 e^- \rightarrow H_2(g) + 2 OH^-(aq)$
  (c) $2 Cl^- \rightarrow Cl_2(g) + 2 e^-$
  (d) $MgCl_2(l) \rightarrow Mg(s) + Cl_2(g)$

18. In the electrolysis of Al_2O_3, the ratio of the masses of Al ( )
and O_2 produced per hour is
  (a) less than one   (b) 2:3
  (c) 1:1   (d) greater than one

19. In purifying copper by electrolysis, which electrode should ( )
be made of pure copper?
  (a) anode   (b) both
  (c) cathode   (d) neither

20. A quantity of 20 000 C is equal to how many moles of ( )
electrons?
  (a) 3×10^{-20} (b) 0.21
  (c) 1.9×10^9 (d) 0.021

21. In the formation of chromium metal from Cr^{3+}, the number ( )
of grams of Cr (AM = 52) produced by one mole of electrons is
  (a) 17   (b) 52
  (c) 104   (d) 156

22. In the reduction of MnO_4^- to Mn^{2+}, the gram equivalent ( )
mass of MnO_4^- is _____ times its gram formula mass.
  (a) 5   (b) 1
  (c) $\frac{1}{3}$ (d) $\frac{1}{5}$

306 • 22–Oxidation and Reduction; Electrochemical Cells

23. When the Zn-Cu^{2+} cell is used to produce electrical energy, ( ) cations
   (a) move toward the Zn electrode, anions to the Cu
   (b) move toward Cu, anions toward Zn
   (c) and anions move toward Zn
   (d) and anions move toward Cu

24. When a lead storage battery is charged, lead sulfate is ( )
   (a) formed at the cathode
   (b) formed at the anode
   (c) formed at both electrodes
   (d) removed from both electrodes

25. Which one of the following reactions could serve as a source ( ) of energy in a fuel cell?
   (a) $H_2O(l) \rightarrow H_2(g) + \frac{1}{2}O_2(g)$
   (b) $Zn(s) + Cu^{2+}(aq) \rightarrow Zn^{2+}(aq) + Cu(s)$
   (c) $CO_2(g) \rightarrow C(s) + O_2(g)$
   (d) $C(s) + O_2(g) \rightarrow CO_2(g)$

26. The carbon atom in formaldehyde, $H-\overset{\overset{O}{\|}}{C}-H$, has an ( ) oxidation number of
   (a) +4
   (b) zero
   (c) -4
   (d) something else

27. In the reaction $CH_3CH_2OH(aq) + O_2(g) \rightarrow CH_3\overset{\overset{O}{\|}}{C}-OH(aq)$ ( ) $+ H_2O$, the species being oxidized is
   (a) CH_3CH_2OH
   (b) O_2
   (c) $CH_3\overset{\overset{O}{\|}}{C}-OH$
   (d) H_2O

28. The oxidation number of nitrogen in hydrazoic acid, HN_3, ( ) is
   (a) -3
   (b) -1
   (c) $-\frac{1}{3}$
   (d) +1

Problems

29. Balance the following equation:

$$P_4(s) \rightarrow PH_3(g) + HPO_3{}^{2-}(aq) \quad \text{(basic solution)}$$

30. When a solution of M^{3+} ions (AM M = 75.0) was electrolyzed, 1.12 g of element M were deposited at the cathode.
    (a) Calculate the number of moles of electrons involved.
    (b) If it took 24 min for the electrolysis, how many amperes were used?

31. A lead storage battery is used to supply a constant current of 10.0 A for 45 min in the electrolytic reduction of $CuBr_2$ to copper.
    (a) Calculate the total charge, in coulombs, "pumped" by the battery.
    (b) Determine the number of grams of copper metal formed (AM Cu = 63.5).

*32. Crystals of silver can be grown by immersing copper wire in silver nitrate solution.
    (a) What is the reaction?
    (b) If a crystal of silver (cubic close-packed, with the shortest internuclear distance of 0.288 nm) is found to grow about 5 mm in length in 30 min, then about how many layers of silver atoms must deposit per second at the surface of the growing crystal?

SELF-TEST ANSWERS

1. T (Its oxidation number thereby decreases.)
2. T (As, for example, in completing an octet. The highest, +5, corresponds to loss of all valence electrons.)
3. F (Such numbers are associated with highly electronegative elements.)
4. F (OH^-.)
5. F
6. T (The melting point is lower for the solution – Chapter 12.)
7. T (Thus maintaining electrical neutrality.)
8. F (To reduce concentration of "free" ion – Chapter 21.)
9. F (One; Ag goes from +1 to 0. Recognizing the charge on a central atom is considered in Chapter 21.)
10. F
11. c (Start by assigning oxygen the usual −2.)
12. d
13. a (No +1 oxidation number.)
14. c
15. d (Balance S; O using H_2O; H using H^+; then charge, adding e^- – in that order.)
16. a (For electrical neutrality, electron loss *must* equal electron gain.)
17. c (In *any* kind of cell, oxidation occurs at the anode.)

308 • 22–Oxidation and Reduction; Electrochemical Cells

18. d (2 × AM of Al: 3 × AM of O – Chapter 3.)
19. c (Cu deposited there.)
20. b (20 000 /96 500.)
21. a (1 mol e$^-$ reacts with $\frac{1}{3}$ mol Cr^{3+}.)
22. d (What is the change in oxidation number?)
23. b (Cu^{2+} is *reduced at the* Cu *cathode*.)
24. d (While in use, i.e., discharging, PbSO$_4$ forms at both.)
25. d (a and c are nonspontaneous under most conditions; what about b?)
26. b (Using the definition of oxidation number, assign bonding electrons to the more electronegative atom – Chapters 7, 8.)
27. a (The oxidation numbers for carbon are, left to right, –3 and –1 in CH$_3$CH$_2$OH; –3 and +3 in CH$_3$CO$_2$H. The equation could partly represent the air oxidation of wine to vinegar.)
28. c
29. $$\frac{\begin{array}{l} P_4(s) + 12\ H^+ + 12\ e^- \rightarrow 4\ PH_3(g) \\ P_4(s) + 12\ H_2O \rightarrow 4\ HPO_3{}^{2-}(aq) + 20\ H^+(aq) + 12\ e^- \end{array}}{2\ P_4(s) + 12\ H_2O \rightarrow 4\ PH_3(g) + 4\ HPO_3{}^{2-}(aq) + 8\ H^+(aq)}$$

or,

$$\frac{\begin{array}{ll} P_4(s) + 6\ H_2O \rightarrow 2\ PH_3(g) + 2\ HPO_3{}^{2-}(aq) + 4\ H^+(aq) & \\ \quad + 4\ OH^-(aq) & + 4\ OH^-(aq) \end{array}}{P_4(s) + 2\ H_2O + 4\ OH^-(aq) \rightarrow 2\ PH_3(g) + 2\ HPO_3{}^{2-}(aq)}$$

30. (a) M^{3+}(aq) + 3 e$^-$ → M(s)

$$1.12\ \text{g M} \times \frac{1\ \text{mol M}}{75.0\ \text{g M}} \times \frac{3\ \text{mol e}^-}{1\ \text{mol M}} = 0.0448\ \text{mol e}^-$$

(b) no. of coulombs = 96 500 (0.0448) = 24 (60) (no. of amperes)

$$\text{no. of amperes} = \frac{96\ 500\ (0.0448)}{1440} = 3.00\ \text{A}$$

31. (a) no. of coulombs = (10.0) (45) (60) = 27 000

(b) $27\ 000\ \text{C} \times \dfrac{1\ \text{mol e}^-}{96\ 500\ \text{C}} \times \dfrac{1\ \text{mol Cu}}{2\ \text{mol e}^-} \times \dfrac{63.5\ \text{g Cu}}{1\ \text{mol Cu}} = 8.88\ \text{g Cu}$

*32. (a) Cu(s) + 2 Ag$^+$(aq) → Cu^{2+}(aq) + 2 Ag(s)

(b) Simplest assumption: imagine silver atoms laid side by side (as along the face diagonal of the unit cell – Chapter 11).

$$\text{rate of growth} = 2.78 \times 10^3\ \text{nm/s} \times \frac{1\ \text{atomic layer}}{0.288\ \text{nm}} = 9650\ \frac{\text{layers}}{\text{s}}$$

(Mind-boggling activity! And those atoms that deposit are only a fraction of those that could.)

SELECTED READINGS

For practice in balancing redox equations, see the problem manuals listed in the Preface.

Alternatives to a fossil fuel economy are discussed in:

Bamberger, C. E., Hydrogen: A Versatile Element, *American Scientist* (July-August 1975), pp. 438–447.

Crowe, B. J., *Fuel Cells: A Survey*, Washington, D. C., NASA, 1973.

Lawrence, R. M., Electrochemical Cells for Space Power, *Journal of Chemical Education* (June 1971), pp. 359–361.

Weissman, E. Y., Batteries: The Workhorses of Chemical Energy Conversion, *Chemistry* (November 1972), pp. 6–11.

23
OXIDATION-REDUCTION REACTIONS; SPONTANEITY AND EXTENT

QUESTIONS TO GUIDE YOUR STUDY

1. How do you know what voltage to apply in carrying out an electrolysis reaction? How do you know what voltage to expect in a voltaic cell?
2. What factors determine the voltage associated with any given redox reaction? (Can you relate the voltage to properties of atoms such as ionization energy and electronegativity?)
3. How can you predict the spontaneity of any given redox reaction? (How have you been able to predict the spontaneity of other reactions?)
4. What factors determine the extent to which a redox reaction proceeds? How can you change the extent of reaction?
5. What quantitative relationships exist between a cell voltage and reaction conditions such as temperature, concentrations and pressure?
6. How might you experimentally show that a given redox reaction is reversible? (Can you think of any common example?)
7. How do you decide what materials may be used for electrodes in voltaic cells? In electrolytic cells?

8. Why does the voltage drop during the use of a voltaic cell? Why do batteries "run down" even when not in use?

9. Are voltaic cells practical major sources of energy? (For example, for lighting and heating a house; for running a car?)

10. To what extent can you convert chemical energy into electrical energy and vice versa?

11.

12.

YOU WILL NEED TO KNOW

Concepts

1. How to recognize oxidation and reduction; how to balance redox equations; and other concepts developed in the preceding chapter.

2. How to interpret the free energy change for a reaction; how to predict the effects of changes in reaction conditions on the free energy change and on the equilibrium constant — Chapters 14, 15.

Math

1. How to use logs and antilogs — Appendix 4.

2. How to work stoichiometric problems for redox reactions — Chapter 22.

3. How to calculate the free energy change for any reaction; the equilibrium constant for the reaction; and how to quantitatively predict the effect of changes in reaction conditions on ΔG and K — Chapters 14, 15.

4. How to calculate K for "multiple equilibria" — see Chapter 15.

CHAPTER SUMMARY

From measurements on voltaic cells, it is possible to obtain numbers called *standard voltages* which are a quantitative measure of the tendency of a species to be reduced or oxidized. A large positive value for the standard reduction voltage implies a species is easily reduced (e.g., $Cl_2(g) + 2\ e^- \rightarrow 2\ Cl^-(aq)$; $E°_{red} = +1.36$ V). Species which are very difficult to reduce have large negative standard reduction voltages ($Al^{3+}(aq) + 3\ e^- \rightarrow Al(s)$; $E°_{red} =$

−1.66 V). Standard oxidation voltages, which can be obtained by changing the sign of the voltage for the reverse reaction, can be interpreted similarly. Species which are difficult to oxidize, such as Cl⁻, have large negative voltages ($E°_{ox}$ Cl⁻ = − 1.36 V); large positive voltages imply a species which is readily oxidized, such as aluminum ($E°_{ox}$ = +1.66 V).

The standard voltage, E°, corresponding to a particular redox reaction can be obtained by adding the standard voltages for the two half-reactions. If the calculated value of E° is positive, we conclude that the reaction is spontaneous at standard concentration and pressure. Such reactions will take place under ordinary laboratory conditions; alternatively, they can serve as a source of electrical energy in a voltaic cell. In contrast, a reaction with a negative E° is nonspontaneous at standard concentration and pressure; it can be carried out only by supplying electrical energy. The electrolyses of Al_2O_3 and NaCl, discussed in Chapter 22, are examples of reactions in this category.

The E° value for a reaction can be directly related to the standard free energy change and hence to the equilibrium constant for the reaction. The relation is:

$$\log_{10} K = \frac{nFE°}{2.30\ RT}; \text{ or, at } 25°C, \log_{10} K = \frac{nE°}{0.0591}$$

where n is the number of moles of electrons transferred in the reaction. From this equation, we see that a redox reaction which has a positive E° value will have an equilibrium constant greater than 1. If E° is negative, K will be less than 1 and the reaction will be nonspontaneous at standard conditions.

Frequently, we need to know the voltage, E, of a cell when reactants and/or products are present at other than standard conditions (1 atm for gases, 1 M for species in aqueous solution). For the general redox reaction:

$$aA + bB \rightarrow cC + dD$$

we write the Nernst equation:

$$E = E° - \frac{0.0591}{n} \log_{10} \frac{(\text{conc. C})^c (\text{conc. D})^d}{(\text{conc. A})^a (\text{conc. B})^b}, \text{ at } 25°C$$

This equation tells us that the voltage drops (E < E°) if the concentration of a product is increased (e.g., conc. C > 1 M); we can increase the voltage of a cell by decreasing the concentration of a product (conc. C < 1 M) or increasing that of a reactant (conc. A > 1 M). The Nernst equation is useful for obtaining concentrations of ions in solution from voltage measurements, particularly with components too dilute to be analyzed for by ordinary chemical methods. The pH meter works on this principle; K_{sp} values

314 • 23–Oxidation-Reduction Reactions; Spontaneity and Extent

(Chapter 18) and K_w, K_a, and K_b for weak acids and bases (Chapter 19) can also be obtained in this way.

In the last three sections of this chapter, we apply the principles reviewed above to organize the descriptive chemistry of redox reactions in water solution. In particular, we examine some of the more important reactions of strong oxidizing agents (species with large positive reduction voltages). Most of these species fall in one of two categories: they are either highly electronegative nonmetals (F_2, Cl_2 O_2), or oxyanions in which the central atom is in its highest oxidation state (MnO_4^-, $Cr_2O_7^{2-}$, NO_3^-). Perhaps the most important redox reaction, at least from an economic standpoint, is the corrosion of iron and steel. Research in this area indicates that corrosion occurs by an electrochemical mechanism. At an anodic area on the surface of an iron object, Fe atoms are oxidized, first to Fe^{2+} and eventually to a product with the approximate composition $Fe(OH)_3$. At the cathode of the tiny voltaic cell, dissolved oxygen is reduced to H_2O molecules or OH^- ions.

BASIC SKILLS

1. Use standard electrode potentials (Table 23.1) to:

    a. compare the relative strengths of different oxidizing agents; different reducing agents.

Referring to Table 23.1, the species in the left column can all, at least in principle, act as oxidizing agents. As one moves down the column, from Li^+ at the top to F_2 at the bottom, the $E°_{red}$ becomes more positive ($E°_{red}$ Li^+ = -3.05 V, F_2 = +2.87 V) and hence oxidizing strength increases. (That is, the species become more easily reduced.)

Species in the right-hand column of Table 23.1 are all potential reducing agents. As one moves up this column, from F^- at the bottom to $Li(s)$ at the top, the $E°_{ox}$ becomes more positive ($E°_{ox}$ F^- = -2.87 V, $Li(s)$ = +3.05 V) and hence reducing strength increases.

See Example 23.3, Problems 23.10 and 23.28.

    b. calculate a cell voltage at standard concentration and pressure.

The relationship here is very simple:

$E° = E°_{red}$ of species which is reduced + $E°_{ox}$ of species which is oxidized.

It can be used to calculate the maximum voltage which is produced by the spontaneous reaction going on in a voltaic cell (Example 23.1), or the minimum voltage which must be applied to carry out a nonspontaneous reaction in an electrolytic cell (Example 23.2).

See Problems 23.1, 23.6, 23.7, 23.24, and 23.25.

c. decide whether or not a given redox reaction will occur spontaneously at standard concentration and pressure.

The principle here is simple. If the calculated $E°$ is positive, the reaction is spontaneous. It would, for example, occur if the species were mixed in a beaker or test tube; alternatively, it could serve as a source of energy in a voltaic cell. If the calculated $E°$ is negative, the reaction is nonspontaneous. Work would have to be done to make the reaction go. It could, for example, be made to take place by supplying electrical energy in an electrolytic cell.

This principle is illustrated in Example 23.4 and applied, in simplest form, in Problems 23.2, 23.11, and 23.29. Three points which you should keep in mind in working problems of this type are:

a. You must combine an oxidation half-reaction with a reduction half-reaction. Students sometimes attempt to combine two oxidations or two reductions, thereby obtaining an absurd answer.
b. Sometimes there will be more than one combination that will give a positive $E°$. When this happens, there will be competing spontaneous reactions; the one which occurs most rapidly (not necessarily the one with the most positive $E°$ value) will predominate.
c. You must use as reactants only those species which are present at relatively high concentrations. For example, in Problem 23.11, it would not be legitimate to write a reaction between Fe^{2+} and Cl_2 (to give Fe^{3+} and Cl^-). Although the calculated $E°$ would be positive, the reaction cannot occur since there is no chlorine gas present.

Still another application of this principle requires that you decide what happens when a given solution is electrolyzed, using the minimum voltage. Here, you can assume that the reaction which has the smallest negative voltage will occur. Problems 23.9 and 23.27 are of this type. In working these problems, don't overlook the possibility of producing H_2 or O_2 from the water present. The appropriate half-reactions are:

$$2 H_2O + 2 e^- \rightarrow H_2(g) + 2 OH^-(aq); E°_{red} = -0.83 \text{ V}$$

$$2 H_2O \rightarrow O_2(g) + 4 H^+(aq) + 4 e^-; E°_{ox} = -1.23 \text{ V}$$

d. calculate $\Delta G°$ for a redox reaction.

The relation here is:

$$\Delta G° \text{ (kJ)} = -96.5nE°$$

316 • 23–Oxidation-Reduction Reactions; Spontaneity and Extent

where E° is the voltage at standard concentration and pressure, as calculated in Skill 1(b), and n is the number of moles of electrons transferred in the reaction as written. Example 23.5 illustrates calculations of this type. See Problems 23.3 and 23.13. Note that in the latter problem, you have to find two half-equations which add to the equation given.

e. calculate the equilibrium constant for a redox reaction.

The equation is:

$$\log_{10} K = nE°/0.0591 \text{ (at } 25°C)$$

where n has the same meaning as in Skill 1d. This calculation is carried out in Example 23.6a. In the latter parts of this example, the applications of K are considered. The reasoning and calculations required there are entirely analogous to those involved in gaseous equilibria, Chapter 15, or acid-base equilibria, Chapter 19.

Of the problems of this type at the end of the chapter, Problems 23.3 and 23.13 are perhaps the simplest. Problems 23.14, 23.15, 23.32, and 23.33 require that you first calculate K from E° and then apply it as in Example 23.6. In Problem 23.31, you are given data from which K can be calculated; this value of K can in turn be used to obtain E° and ΔG°.

2. Use the Nernst equation (Equation 23.6) to calculate:

a. the voltage of a cell, given E° and the concentrations of all species.

A typical calculation of this sort is shown in Example 23.7. Again, in Example 23.9, the Nernst equation is applied, this time to determine the effect of concentration upon the voltage of a half cell. See also Problems 23.4(a), 23.16, 23.17, 23.34, and 23.35.

b. the concentration of one species, given those of all other species, E, and E°.

See Example 23.8 and Problem 23.4(b). By an extension of this approach, cell voltages can be combined with measured concentrations to determine equilibrium constants for various kinds of solution reactions.

Referring to Example 23.8 in the text, suppose the source of H⁺ ions in the H⁺-H₂ half-cell is a 0.10 M solution of the weak acid HCN. Calculate the ionization constant for HCN. _____

In Example 23.8, we concluded that $[H^+] = 8 \times 10^{-6}$ M. Since the H^+ ion comes from the dissociation of 0.10 M HCN:

$$HCN(aq) \rightarrow H^+(aq) + CN^-(aq)$$

it follows that $[CN^-] = [H^+] = 8 \times 10^{-6}$ M and $[HCN] = 0.10$ M. Therefore:

$$K_a = \frac{[H^+][CN^-]}{[HCN]} = \frac{(8 \times 10^{-6})(8 \times 10^{-6})}{(1 \times 10^{-1})} = 6 \times 10^{-10}$$

Problems 23.19 and 23.37 illustrate qualitatively the principle just illustrated. In Problems 23.18 and 23.36, you must use this skill to determine equilibrium constants for precipitation (K_{sp}) and complex ion dissociation (K_d), respectively.

3. **Given a balanced equation for a redox reaction and titration data for the reaction, calculate the concentration of one of the reactant species.**

See Example 23.10. No new principle is introduced here. The approach followed is entirely analogous to that used with precipitation reactions (Chapter 18) and acid-base reactions (Chapter 20). Problems 23.5, 23.23, and 23.41 are entirely analogous to Example 23.10.

In addition to these specific skills, a considerable amount of descriptive chemistry dealing with redox reactions is presented in Sections 23.4, 23.5, and 23.6. You should be familiar with:

a. the reaction of Cl_2 with Br^- and I^- ions.
b. methods of preparation of the oxyanions of chlorine.
c. the oxidation numbers of nitrogen and chromium (Table 23.3).
d. the mechanism of corrosion of iron.
e. the use of strong oxidizing agents to bring insoluble solids into solution in qualitative analysis.

See Problems 23.20, 23.21, 23.22, 23.38, 23.39, and 23.40.

SELF-TEST

True or False

1. The strongest oxidizing agents have the largest, most ( ) positive $E°_{red}$.

318 • 23–Oxidation-Reduction Reactions; Spontaneity and Extent

2. The standard reduction voltages of F_2 and Ag^+ are +2.87 ( )
V and +0.80 V, respectively. We conclude that F^- is a better reducing agent than Ag metal.

3. For a certain mixture, positive E° values are calculated for ( )
two different redox reactions. We can be confident that the one with the higher E° value will occur first.

4. The voltage of a cell in which the reaction $Cu(s) +$ ( )
$2\ Ag^+(aq) \rightarrow Cu^{2+}(aq) + 2\ Ag(s)$ occurs will be, at standard concentrations, $E°_{ox} Cu + 2 \times E°_{red} Ag^+$.

5. Reactions which are readily reversed by a small change in ( )
concentration are those in which E° is close to zero.

6. For the cell referred to in Question 4, increasing the ( )
concentration of Ag^+ by a factor of 10 will increase the voltage by +0.06 V.

7. The oxidizing strength of oxyanions is ordinarily greatest at ( )
low pH.

8. The stable species of the element nitrogen with a +3 ( )
oxidation number in acidic solution is HNO_3.

9. The phrase "cathodic protection" refers to the common ( )
practice of enclosing fragile metal electrodes in Plexiglas to prevent them from being broken.

10. The number of electrons transferred in the reaction $2\ BrO_3^-$ ( )
$+ 3\ N_2H_4 \rightarrow 2\ Br^- + 3\ N_2 + 6\ H_2O$ is six.

Multiple Choice

11. Referring to Table 23.1, if the standard reduction voltage ( )
of Ni^{2+} were set at 0.00 V, that of Mg^{2+} would be
    (a) -2.12 V   (b) +2.12 V
    (c) -2.62 V   (d) +2.62 V

12. The standard reduction voltages of Cl_2 and Cu^{2+} are +1.36 ( )
V and +0.34 V, respectively. The E° value for the reaction $Cu^{2+}(aq) + 2\ Cl^-(aq) \rightarrow Cu(s) + Cl_2(g)$ is
    (a) -2.38 V   (b) -1.70 V
    (c) -1.02 V   (d) +1.70 V

13. The E° values for the following reactions are known to be ( )
positive:

$$A(s) + B^{2+}(aq) \rightarrow A^{2+}(aq) + B(s)$$

$$A(s) + C^{2+}(aq) \rightarrow A^{2+}(aq) + C(s)$$

At standard concentrations, the reaction between B^{2+} and C
- (a) is spontaneous
- (b) is nonspontaneous
- (c) is at equilibrium
- (d) cannot say

14. Given the following standard reduction voltages: ( )

$Mn^{2+}(aq) + 2 e^- \rightarrow Mn(s)$ -1.18 V
$2 H_2O + 2 e^- \rightarrow H_2(g) + 2 OH^-(aq)$ -0.83 V
$I_2(s) + 2 e^- \rightarrow 2 I^-(aq)$ $+0.53$ V
$O_2(g) + 4 H^+(aq) + 4 e^- \rightarrow 2 H_2O$ $+1.23$ V

we would predict that the electrolysis of a water solution of MnI_2 would probably produce
- (a) Mn, I_2
- (b) Mn, O_2
- (c) H_2, I_2
- (d) H_2, O_2

15. For a certain redox reaction, $E°$ is positive. This means that ( )
- (a) $\Delta G°$ is positive, K is greater than 1
- (b) $\Delta G°$ is positive, K is less than 1
- (c) $\Delta G°$ is negative, K is greater than 1
- (d) $\Delta G°$ is negative, K is less than 1

16. For the reaction $4 Al(s) + 3 O_2(g) + 6 H_2O \rightarrow 4 Al(OH)_3(s)$, ( ) n in the equation $\Delta G° = -nFE°$ is
- (a) 1
- (b) 2
- (c) 3
- (d) 12

17. For the redox reaction $A(s) + B^{2+}(aq) \rightleftarrows A^{2+}(aq) + B(s)$, K = ( ) 10. When the concentrations of B^{2+} and A^{2+} are 0.5 M and 0.1 M, respectively,
- (a) the forward reaction is spontaneous
- (b) the system is at equilibrium
- (c) the reverse reaction is spontaneous
- (d) cannot say

18. It is possible to increase the voltage of a cell in which the ( ) reaction is $Zn(s) + Cu^{2+}(aq) \rightarrow Zn^{2+}(aq) + Cu(s)$ by increasing the
- (a) concentration of Zn^{2+}
- (b) concentration of Cu^{2+}
- (c) size of the Zn electrode
- (d) size of the Cu electrode

19. Which one of the following equilibrium constants would be ( ) most difficult to obtain from cell measurements?
- (a) K_b for NH_3
- (b) K_c for $H_2(g) + Cl_2(g) \rightarrow 2 HCl(g)$
- (c) K_d for $Cu(NH_3)_4^{2+}$
- (d) K_w for water

320 • 23–Oxidation-Reduction Reactions; Spontaneity and Extent

20. A solution of NaOH saturated with Cl_2 at room temperature will contain appreciable concentrations of all but one of the following species. Indicate the exception. ( )
    (a) Cl^- (b) OH^-
    (c) ClO^- (d) ClO_4^-

21. Which one of the following metals reacts with HNO_3 but not with dilute HCl? ( )
    (a) Pt   (b) Mg
    (c) Na   (d) Cu

22. Which one of the following metals reacts with dilute HCl but not with water? ( )
    (a) Ag   (b) Na
    (c) Ni   (d) Ca

23. In order to convert CrO_4^{2-} to $Cr_2O_7^{2-}$, one would add ( )
    (a) water   (b) an acid
    (c) an oxidizing agent   (d) a reducing agent

24. Corrosion ordinarily occurs more readily in seawater than in fresh water because ( )
    (a) Na^+ ions in seawater attack iron
    (b) Cl^- ions in seawater attack iron
    (c) seawater is a better electrical conductor
    (d) O_2 is more soluble in seawater

25. "Aqua regia," a mixture of concentrated HCl and concentrated HNO_3, dissolves certain metals and metal sulfides which fail to dissolve in concentrated HNO_3 alone. The main function of the HCl is to ( )
    (a) increase the H^+ concentration
    (b) furnish Cl_2, a better oxidizing agent than HNO_3
    (c) furnish Cl^-, which acts as a complexing agent
    (d) convert the metal to gold, which is soluble in HNO_3

26. Reactions which could occur, at least in principle, in a voltaic cell include ( )
    (a) $Ag(s) + \frac{1}{2} Cl_2(g) \rightarrow AgCl(s)$
    (b) $Ag^+(aq) + Cl^-(aq) \rightarrow AgCl(s)$
    (c) $AgCl(s) + 2 NH_3(aq) \rightarrow Ag(NH_3)_2^+(aq) + Cl^-(aq)$
    (d) all of these

27. Using Table 23.1, $E°$ for the reaction $MgCl_2(l) \rightarrow Mg(s) + Cl_2(g)$ ( )
    (a) is found to be +3.73 V   (b) is found to be −3.73 V
    (c) is found to be +1.01 V   (d) cannot be determined

28. Electrolysis of very dilute, say 0.001 M, aqueous solutions ( )
of $CuBr_2$ is expected to give the products
 (a) Cu, Br_2 (b) Cu, O_2
 (c) H_2, O_2 (d) H_2, Br_2

29. In the fuel cell employing the reaction $CH_4 + 2 O_2 \rightarrow CO_2 +$ ( )
2 H_2O,
 (a) carbon is reduced
 (b) hydrogen is oxidized
 (c) CH_4 and O_2 react at the same electrode
 (d) chemical energy is converted directly to electrical energy

30. The voltaic cell is named after ( )
 (a) the electrical unit, volt
 (b) the Italian scientist, Alessandro Volta
 (c) the Volta River, in Africa, infested with electric eels
 (d) the last name of the Russian scientist, Sergei Gregorovich Atlov, spelled backwards

Problems

31. Given:

$Fe^{3+}(aq) + e^- \rightarrow Fe^{2+}(aq); E°_{red} = +0.77$ V

$Fe^{2+}(aq) + 2 e^- \rightarrow Fe(s); E°_{red} = -0.44$ V

Calculate:

 (a) $E°$ for the disproportionation reaction $3 Fe^{2+}(aq) \rightarrow 2 Fe^{3+}(aq) + Fe(s)$.
 (b) K for the reaction in (a).
 (c) E for the reaction $3 Fe^{2+}(aq, 0.010$ M$) \rightarrow 2 Fe^{3+}(aq, 0.010$ M$) + Fe(s)$

32. Given the reaction $4 Cr^{2+}(aq) + O_2(g) + 4 H^+(aq) \rightarrow 4 Cr^{3+}(aq) + 2 H_2O$ and the data:

$Cr^{3+}(aq) \rightarrow Cr^{2+}(aq), E°_{red} = -0.41$ V
$O_2(g) \rightarrow H_2O, \quad E°_{red} = +1.23$ V

 (a) Write the half-equation for the reduction and assign $E°_{red}$.
 (b) Write the half-equation for the oxidation and assign $E°_{ox}$.
 (c) Calculate $E°$ for the reaction.

(d) Calculate $\Delta G°$ for the reaction.
(e) Calculate K for the reaction.
(f) Would you expect acidic solutions of Cr^{2+} in air to be stable?

33. The standard free energy of formation of AgBr(s) is -95.9 kJ/mol.

(a) Calculate the standard voltage of a cell with the overall reaction

$$Ag(s) + \frac{1}{2} Br_2(l) \rightarrow AgBr(s)$$

(b) The standard reduction voltage for the half-reaction

$$Br_2(l) + 2\ e^- \rightarrow 2\ Br^-(aq)$$

is +1.07 V. Calculate the standard reduction voltage for

$$AgBr(s) + e^- \rightarrow Ag(s) + Br^-(aq)$$

*34. Derive an equation that relates the cell voltage to the pH in the half-cell, operating at 25°C and one atmosphere, $O_2(g) + 4\ H^+(aq) + 4\ e^- \rightarrow 2\ H_2O$.

SELF-TEST ANSWERS

1. T (Most easily reduced.)
2. F (The reverse is true.)
3. F (Will depend on relative rates.)
4. F ($E°_{ox}$ Cu + $E°_{red}$ Ag^+.)
5. T
6. T (Write the Nernst equation.)
7. T (High $[H^+]$ favors their reduction; apply Le Chatelier's principle – Chapter 15.)
8. F (HNO_2, and not NO_2^-, since it is a weak base – Chapter 19.)
9. F
10. F (Twelve: 2 bromine atoms go from O.N. = +5 to -1. On how to assign oxidation numbers, see Chapter 22.)
11. a (Their *relative* tendencies for reduction remain unchanged.)
12. c
13. d ($E° > 0$ for first equation; $E° < 0$ for reversed second; their sum = ?)
14. c (Predominant species are Mn^{2+}, I^-, and H_2O. Look for least negative sum: $E°_{ox} + E°_{red}$.)

Self-Test Answers • 323

15. c
16. d (Reasoning outlined in answer to 10, above.)
17. a (Compare concentration ratio to K — Chapter 15.)
18. b (Think like Le Chatelier — Chapter 15.)
19. b (Involves three gases!)
20. d (Halogens tend to disproportionate in basic solution; see the text discussion.)
21. d (Determine E° for the possible reactions; NO_3^- rather than H^+ is oxidizing agent.)
22. c (H^+, not H_2O, is reduced by Ni.)
23. b (Can you write the equation? What are the oxidation numbers?)
24. c
25. c (Complexing the product cation effectively removes it, shifting the position of equilibrium, hence making E more positive.)
26. d (All are spontaneous under standard conditions. Recall definition of spontaneity, Chapter 14.)
27. d (Molten, not aqueous, reactant! The table applies *only* to aqueous solution.)
28. c (Predominant species now only H_2O.)
29. d
30. b (With thanks to Professor S. Ruven Smith.)
31. (a) $2(Fe^{2+}(aq) \rightarrow Fe^{3+}(aq) + e^-)$ $E°_{ox} = -0.77$ V
     $Fe^{2+}(aq) + 2e^- \rightarrow Fe(s)$ $E°_{red} = -0.44$ V
     ─────────────────────────────────────
     $3 Fe^{2+}(aq) \rightarrow 2 Fe^{3+}(aq) + Fe(s)$ $E° = -1.21$ V

   (b) $\log K = \dfrac{2(-1.21)}{0.0591} = -40.9;\ K = 1 \times 10^{-41}$

   (c) $E = -1.21 - \dfrac{0.0591}{2} \log \dfrac{(0.010)^2}{(0.010)^3} = -1.21 - 0.06 = -1.27$ V

32. (a) $\frac{1}{2} O_2(g) + 2 H^+(aq) + 2 e^- \rightarrow H_2O$; $E°_{red} = +1.23$ V
    (b) $Cr^{2+}(aq) \rightarrow Cr^{3+}(aq) + e^-$; $E°_{ox} = +0.41$ V
    (c) $E° = +1.23\text{ V} + 0.41\text{ V} = +1.64$ V
    (d) $\Delta G° = -96.5(4)(+1.64) = -633$ kJ
    (e) $\log K = \dfrac{4(1.64)}{0.0591} = 111;\ K = 10^{111}$
    (f) No; though nothing can be said about reaction rate.

33. (a) $Ag(s) + Br^-(aq) \rightarrow AgBr(s) + e^-$
     $\frac{1}{2} Br_2(l) + e^- \rightarrow Br^-(aq)$
     ─────────────────────────────────────
     $Ag(s) + \frac{1}{2} Br_2(l) \rightarrow AgBr(s)$
     $-95.9 = -96.5(1)E°;\ E° = +0.994$ V

   (b) $1.07\text{ V} + E°_{ox}\ Ag = +0.99$ V; $E°_{ox}\ Ag = -0.08$ V
                                                    $E°_{red}\ Ag = +0.08$ V

*34. For P_{O_2} = 1 atm, T = 298 K:

$$E = E° - \frac{0.0591}{4} \log \frac{1}{[H^+]^4} = E° - \frac{0.0591}{4}(-4) \log [H^+]$$

$$E = 1.23 - 0.0591 \text{ pH}$$

SELECTED READINGS

See the Readings listed for the preceding chapter, and:

Fischer, R. B., Ion-Selective Electrodes, *Journal of Chemical Education* (June 1974), pp. 387–390.

Slabaugh, W. H., Corrosion, *Journal of Chemical Education* (April 1974), pp. 218–220.

Taube, H., Mechanisms of Oxidation-Reduction Reactions, *Journal of Chemical Education* (July 1968), pp. 452–461.

24
NUCLEAR REACTIONS

QUESTIONS TO GUIDE YOUR STUDY

1. How are nuclear reactions different from "ordinary" chemical reactions? How would you experimentally recognize that a particular reaction was a nuclear reaction?

2. What factors determine whether a particular atom is radioactive? Are there correlations that can be made with the Periodic Table? With nuclear composition?

3. What are the properties of the various kinds of radiation? How, for example, do they interact with matter? What chemical reactions occur as a result of interaction with biological systems?

4. How do reaction conditions such as temperature, pressure and concentration affect the nature of a nuclear reaction?

5. How would you experimentally determine the rate of a nuclear reaction? What can you say about the rate law and reaction mechanism for a given nuclear reaction?

6. What can you say about the spontaneity and extent of a nuclear reaction?

7. What is the difference between "natural" and "artificial" nuclear reactions? Between fission and fusion?

8. What are some of the *chemical* applications of nuclear reactions?

9. What energy effects are associated with nuclear reactions? Can you predict whether a given nuclear reaction will be exothermic or endothermic?

10. What reactions occur in a nuclear power plant? What are some of the advantages and disadvantages of nuclear power? (How, for example, does the cost compare with that of conventional power from fossil fuels?)

11.

12.

YOU WILL NEED TO KNOW

Concepts

1. How to symbolize (and describe) nuclear composition; i.e., how to use the notation that shows the atomic number and the mass number — Chapter 2.
2. The meaning of activation energy, E_a — Chapter 16.

Math

1. How to work problems involving first order kinetics — Chapter 16.
2. How to relate molecular speed and kinetic energy to temperature — Chapter 5.

CHAPTER SUMMARY

In this chapter, we considered three different kinds of spontaneous nuclear reactions.

1. *Radioactive decay*, in which an unstable nucleus decomposes, emitting

$$\text{an alpha particle: } {}^{238}_{92}U \rightarrow {}^{4}_{2}He + {}^{234}_{90}Th$$

$$\text{a beta particle: } {}^{234}_{90}Th \rightarrow {}^{0}_{-1}e + {}^{234}_{91}Pa$$

$$\text{or a positron: } {}^{30}_{15}P \rightarrow {}^{0}_{1}e + {}^{30}_{14}Si$$

and producing a new, more stable nucleus. Several steps may be required to form a nonradioactive nucleus. The decomposition of uranium-238 passes through 14 intermediates (8 α emissions, 6 β) yielding, as a final product, a stable isotope of lead, ${}^{206}_{82}Pb$.

A few radioactive isotopes, mostly those of the heavy elements, occur in nature. Many others have been produced in the laboratory by bombardment reactions using positively charged particles (protons, deuterons, α-particles) or neutrons. An important accomplishment in this area has been the synthesis of at least one isotope each of elements of atomic number 96 to 106, the transuranium elements.

Radioactive decay follows the first order rate law (Chapter 16). Rates of different decay processes are often compared by citing half-lives, which can vary from a millisecond to many billions of years. Methods based on the decay of naturally occurring radioactive isotopes have been worked out to determine the age of rocks, both lunar and terrestrial, and carbon-containing artifacts.

2. *Fission*, in which a heavy nucleus, commonly $^{235}_{92}U$ or $^{239}_{94}Pu$, splits under neutron bombardment to give two lighter isotopes. A typical reaction is:

$$^{235}_{92}U + ^{1}_{0}n \rightarrow ^{90}_{37}Rb + ^{144}_{55}Cs + 2^{1}_{0}n$$

Fission ordinarily produces an excess of neutrons; provided a certain critical mass of fissionable material is present, a chain reaction can result.

3. *Fusion*, in which two light nuclei combine, e.g.,

$$^{2}_{1}H + ^{2}_{1}H \rightarrow ^{4}_{2}He$$

Processes of this type, unlike other types of nuclear reactions, have large activation energies. They occur at reasonable rates only at very high temperatures. As of this writing, the only feasible way to achieve and maintain these temperatures is by means of a fission reaction.

All spontaneous nuclear reactions evolve large amounts of energy, most of it in the form of heat. The quantity of energy given off per unit mass of reactant increases in the order: radioactive decay $<<$ fission $<$ fusion. The energy change can be accounted for quantitatively by the Einstein relation:

$$\Delta E(kJ) = 9.00 \times 10^{10} \times \Delta m (g)$$

Here, unlike ordinary chemical reactions, there is a detectable difference in mass between products and reactants. The enormous amounts of energy evolved in fission and fusion reflect the fact that the binding energy per gram is a maximum for isotopes of intermediate mass (Fig. 24.5). Since the plot of binding energy per gram vs. mass number rises very steeply near the origin and falls off more gradually at high mass numbers, considerably more energy is given off in fusion than in fission.

The reactions discussed in this chapter, none of which were known or even suspected a century ago, have had a profound effect upon our lives and our environment. A generation, born since the holocausts of Hiroshima and Nagasaki, has lived with the constant threat of a nuclear disaster that could destroy life on earth. Now, facing an energy crisis, we are becoming aware of the promise of nuclear reactions, particularly those of the fusion type, to supplement and eventually replace our dwindling supply of fossil fuels.

BASIC SKILLS

1. Write a balanced equation for a nuclear reaction, given the identities of all but one of the reactants and products.

The basic principle here is that both the total mass number (superscript at upper left) and the total nuclear charge (subscript at lower left) must "balance," i.e., must have the same value on both sides of the equation. This principle is illustrated in Examples 24.1 and 24.2.

Problems 24.1, 24.6, 24.8, 24.22, and 24.24 illustrate this skill further. Problems 24.7 and 24.23 are perhaps the most difficult of this type but are readily solved when you realize that the entire mass number change must be accounted for by the α-particles given off, since a β-particle has a mass number of zero.

2. Use the first order rate law and the expression for the half-life to relate the amount of a radioactive species to elapsed time.

The pertinent equations here are those discussed in Chapter 16 in connection with first order reactions:

$$\log_{10} \frac{X_0}{X} = \frac{kt}{2.30}; \quad t_{1/2} = \frac{0.693}{k}$$

Examples 24.3 and 24.4 illustrate the application of these equations to radioactive processes. Problems 24.2, 24.9–24.12 and 24.25–24.28 can be worked in much the same way. In Problem 24.26, note that a plot of log X vs. t should give a straight line for a first order reaction. You should realize that no new concepts are introduced here; very similar calculations were carried out in the problems at the end of Chapter 16.

3. Given a table of nuclear masses such as Table 24.3, calculate for a given isotope the mass decrement (g/mol) and binding energy (kJ/mol or kJ/g).

— —

What is the mass decrement for $^{28}_{13}Al$?_____ The binding energy in kJ/mol?_____ The binding energy in kJ/g?_____

The mass decrement is the decrease in mass when one mole of Al-28 is formed from protons and neutrons. From Table 24.3:

mass 13 mol protons = 13(1.007 28)g
mass 15 mol neutrons = 15(1.008 67)g
_____
                       28.224 69 g

mass one mole of $^{28}_{13}$Al = $\dfrac{\begin{array}{r}28.224\ 69\text{ g}\\ 27.974\ 77\text{ g}\end{array}}{0.249\ 92\text{ g}}$ = mass decrement

To obtain the binding energy in kJ/mol, we use the conversion factor
9.00×10^{10} kJ = 1 g

binding energy (kJ/mol) = $0.249\ 92 \dfrac{\text{g}}{\text{mol}} \times 9.00 \times 10^{10} \dfrac{\text{kJ}}{\text{g}}$

= 2.25×10^{10} kJ/mol

To convert from kJ/mol to kJ/g, we note that one mole of Al-28 weighs 28.0 g.

binding energy (kJ/g) = $2.25 \times 10^{10} \dfrac{\text{kJ}}{\text{mol}} \times \dfrac{1\text{ mol}}{28.0\text{ g}} = 8.04 \times 10^{8}$ kJ/g

---

See Problems 24.18 and 24.34.

4. Using Table 24.3, calculate Δm for a nuclear reaction and relate it to the energy change, ΔE.

The change in mass is readily obtained as in Skill 3. Note that the use of Table 24.3 leads directly to Δm in grams per mole of reactant. Thus, for

$$^{239}_{94}\text{Pu} \rightarrow {}^{235}_{92}\text{U} + {}^{4}_{2}\text{He}$$

we have

$\Delta m = 234.9934\text{ g} + 4.0015\text{ g} - 239.0006\text{ g} = -0.0057\text{ g}$

This difference in mass is readily converted to an energy difference using the conversion factor referred to previously:

$\Delta E = -0.0057\text{ g} \times 9.00 \times 10^{10} \dfrac{\text{kJ}}{\text{g}} = -5.1 \times 10^{8}$ kJ

Note that:
- this calculation gives Δm or ΔE *per mole* of reactant; if you want the change per gram of reactant, you must divide by the molar mass.
- a negative Δm and ΔE imply a spontaneous nuclear reaction; if Δm and ΔE are positive, the reaction is nonspontaneous.

Calculations of this type are carried out in Examples 24.5 and 24.6; they are required in Problems 24.3, 24.14, 24.17, 24.19, 24.30, 24.33, 24.35, and, less obviously, in 24.16 and 24.32.

SELF-TEST

True or False

1. The emission of a β-particle leaves the atomic number ( ) unchanged but increases the mass number by one unit.

2. The extent of deflection in an electric field is greater for a ( ) β-particle than for an α-particle.

3. Emission of a positron is equivalent to the conversion of a ( ) proton to a neutron in the nucleus.

4. The most serious effect of long-term low-level radiation on ( ) the body is that it produces severe skin burns.

5. The longer the half-life of a radioactive isotope, the more ( ) rapidly it decays.

6. The very heavy transuranium elements, atomic number 101 ( ) or greater, are most readily prepared by neutron bombardment.

7. Our ability to detect and measure radiation is much greater ( ) than for most other hazards.

8. According to the Einstein relation (Equation 24.17), the ( ) fusion of one gram of deuterium would liberate 9.00×10^{10} kJ of energy.

9. Probably the most important hazard associated with a ( ) nuclear power plant is the possibility of a *nuclear* explosion.

10. A plausible way to achieve the high temperatures required ( ) for nuclear fusion is to operate the process in the upper atmosphere.

Multiple Choice

11. The emission of an α-particle lowers the atomic number by ( ) _____ and the mass number by _____, respectively.
    (a) 1, 1   (b) 1, 2
    (c) 2, 2   (d) 2, 4

12. Nuclear reactions differ from ordinary chemical reactions in all but one of the following ways. Indicate the exception.
    (a) The energy evolved per gram is much greater for nuclear reactions.
    (b) Nuclear reactions occur much more rapidly.

(c) New elements are often formed in nuclear reactions.
(d) In nuclear reactions, reactivity is essentially independent of the state of chemical combination.

13. Emission of which one of the following leaves both atomic ( ) number and mass number unchanged?
    (a) positron
    (b) neutron
    (c) α-particle
    (d) γ-radiation

14. A certain radioactive series starts with $^{235}_{92}U$ and ends with ( ) $^{207}_{82}Pb$. In the overall process, _____ α-particles and _____ β-particles are emitted.
    (a) 8, 6
    (b) 14, 10
    (c) 7, 10
    (d) 7, 4

15. Which one of the following instruments would be least ( ) suitable for detecting particles given off in radioactive decay?
    (a) Geiger counter
    (b) scintillation counter
    (c) electron microscope
    (d) cloud chamber

16. In determining the age of organic material, one measures ( )
    (a) the time required for half of the C-14 in the sample to decay
    (b) the ratio of C-14 to C-12 in the sample
    (c) the percentage of carbon in the sample
    (d) the time required for half the organic material to decay

17. The half-life of uranium-238, which decomposes to lead-206, is about 4.5×10^9 a. A rock which contains equal numbers of grams of these two isotopes would be _____ years old.
    (a) less than 4.5×10^9
    (b) 4.5×10^9
    (c) more than 4.5×10^9
    (d) cannot say

18. Bombardment of $^{75}_{33}As$ by a deuteron, ^2_1H, forms a proton ( ) and an isotope which has a mass number of _____ and an atomic number of _____.
    (a) 73, 32
    (b) 75, 33
    (c) 75, 32
    (d) 76, 33

19. The element silicon has an atomic mass of about 28. The ( ) isotope $^{30}_{14}Si$ would most likely decay by emitting a(n)
    (a) positron
    (b) protron
    (c) electron
    (d) alpha particle

20. Which one of the following isotopes would be most likely to ( ) undergo fission?
    (a) $^{14}_6C$
    (b) $^{59}_{27}Co$
    (c) $^{239}_{94}Pu$
    (d) ^2_1H

21. Which of the isotopes listed in Question 20 would you ( )
expect to have the largest binding energy per gram?

22. Which of the isotopes listed in Question 20 would produce ( )
the most energy during fusion with an identical nucleus?

23. According to the Einstein relation (Equation 24.17), the ( )
energy given off in a nuclear reaction in which the decrease in mass is
2.0 mg would be
   (a) 1.8×10^{11} kJ   (b) 9.0×10^{10} kJ
   (c) 1.5×10^{10} kJ   (d) some other amount

24. The masses of $^4_2$He, $^6_3$Li, and $^{10}_5$B are 4.0015, 6.0135, and ( )
10.0102, respectively. The splitting of a boron-10 nucleus to
helium-4 and lithium-6 would
   (a) evolve energy
   (b) absorb energy
   (c) result in no energy change
   (d) cannot say

25. About 99% of naturally occurring carbon atoms are the ( )
stable $^{12}_6$C isotope. What type of radioactivity would you expect
from the unstable isotope $^{11}_6$C?
   (a) $_{-1}^{0}$e   (b) $_{+1}^{0}$e
   (c) $^4_2$He   (d) n

26. The decay of a neutron to a proton also yields a(n) ( )
   (a) $_{-1}^{0}$e   (b) $_{+1}^{0}$e
   (c) α-particle   (d) $^2_1$H$^+$

27. The half-lives of radioactive $^{235}_{92}$U and $^{238}_{92}$U are 0.71×10^9 ( )
and 4.51×10^9 a, respectively. For separate samples of these two
isotopes containing equal numbers of atoms, the rate of decay is
more rapid in
   (a) ^{235}U   (b) ^{238}U
   (c) both decay at the same rate

28. Which of the following radioactive isotopes would you use ( )
to date an object containing each one of them if the object is
expected to be about a hundred years old?
   (a) ^{87}Rb ($t_{1/2} = 5.7 \times 10^{10}$ a)
   (b) ^{14}C ($t_{1/2} = 5720$ a)
   (c) ^{63}Ni ($t_{1/2} = 92$ a)
   (d) ^{3}H ($t_{1/2} = 12.3$ a)

29. To estimate the age of the universe, you might use ( )
   (a) $^{187}_{75}$Re → $^{187}_{76}$Os + $_{-1}^{0}$e ($t_{1/2} = 4.4 \times 10^{10}$ a)
   (b) ^{14}C ($t_{1/2} = 5720$ a)

(c) a sundial
(d) an astrologer

30. The treatment of poisoning by ingested heavy metals that ( )
are radioactive is likely to involve the use of
    (a) a chelating agent   (b) a reducing agent
    (c) x-ray therapy   (d) aqua regia

31. For the solar reaction $4\,{}^{1}_{1}H \rightarrow {}^{4}_{2}He + 2{}^{0}_{1}e$, $\Delta E = -6.0 \times 10^8$ ( )
kJ/g H. For each mole of hydrogen undergoing fusion,
    (a) the sun radiates 6.0×10^8 kJ to the surroundings
    (b) the mass of the sun decreases by 1 g
    (c) the mass of the sun increases by 1 g
    (d) the sun absorbs 6.0×10^8 kJ

32. Which of the following statements is not true?   ( )
    (a) Some harmful effects of radioactivity are the result of ionization.
    (b) One advantage of generating electric power in nuclear reactors is the small amount of wasted heat.
    (c) Fusion reactions tend to have very high activation energies.
    (d) Several breeder reactors are already in operation.

Problems

33. A piece of wood found in a cave has a ${}^{14}C/{}^{12}C$ ratio of 5.75×10^{-13}; a piece of wood growing in the same area today has a ratio of 7.42×10^{-13}. If the half-life of ${}^{14}C$ is 5720 a, how old is the wood found in the cave?

34. Nuclear reactions occurring in stars are thought to include the fusion of helium nuclei to form such nuclei as ${}^{8}_{4}Be$, ${}^{12}_{6}C$ and ${}^{16}_{8}O$.
    (a) Write a balanced equation for the fusion of ${}^{4}_{2}He$ nuclei to give a ${}^{12}_{6}C$ nucleus. Show all atomic and mass numbers as well as the charges on these species.
    (b) Calculate ΔE (kJ) for the formation of a mole of ${}^{12}_{6}C$ nuclei by this reaction. (atomic masses: ${}^{4}_{2}He$ = 4.001 50, ${}^{12}_{6}C$ = 11.996 71; 1 g mass = 9.00×10^{10} kJ)

35. Consider the decay of ${}^{210}_{84}Po$, which gives off an alpha particle.
    (a) Write a nuclear equation for the decay.
    (b) If the half-life for the decay is 140 d, what is the rate constant?
    (c) How long will it take for 90% of a sample of Po to be converted to lead?

334 • 24–Nuclear Reactions

*36. Determine whether or not the number of naturally occurring radioactive isotopes is a periodic function of atomic number.

*37. The modern German and English languages are descendents of the Germanic language that split in two directions about 500 A.D. There is evidence that fundamental vocabularies change at such a rate that about 19% of the terms are replaced every thousand years. Estimate the percentage of the vocabularies that remain essentially unchanged in these two languages today. Describe an experiment *involving nuclear reactions* that would verify the date of 500 A.D. (more or less).

SELF-TEST ANSWERS

1. F (Reverse is true.)
2. T (Much smaller mass more than compensates for smaller charge.)
3. T (Mass number and charge balance.)
4. F (Many early workers in the field developed cancer.)
5. F ($k \propto 1/t_{1/2}$.)
6. F (Heavier particles used.)
7. T (See the Readings.)
8. F (Δm must be one gram.)
9. F (Controversial among *non*scientists! See Readings and recent press reports.)
10. F (Concentration of high-temperature, i.e., high-speed, particles is too low.)
11. d
12. b (*Some* are very slow — consider the range of half-lives for decay.)
13. d (Nucleus drops to a lower energy level: see Readings.)
14. d (Total mass change is due to seven α-particles.)
15. c
16. b
17. c (Contains more moles of Pb. This assumes all the Pb comes from U.)
18. d (Write the balanced equation.)
19. c (To give $^{30}_{15}P$ with a mass number more nearly equal to the AM.)
20. c
21. b (See Figure 24.5.)
22. d (Why?)
23. d (1.8×10^8 kJ.)
24. b (Δm is positive.)
25. b (Same kind of reasoning as for Question 19.)
26. a (Write the balanced equation.)

Selected Readings • 335

27. a (See Question 5.)
28. c (So that ratio of reactant to product is easily measured, neither very large nor very small.)
29. a (A recent estimate gave 2×10^{10} a! See *Chemical and Engineering News*, July 12, 1976.)
30. a (To form a very stable, soluble complex – Chapter 21.)
31. a (While its mass decreases by $6.0 \times 10^8 / 9.00 \times 10^{10}$ g.)
32. b (Thermal pollution is a serious disadvantage; see Readings.)
33. $k = \dfrac{0.693}{5720 \text{ a}} = 1.21 \times 10^{-4} \text{ a}^{-1}$

   $\log \dfrac{7.42}{5.75} = 0.111 = \dfrac{1.21 \times 10^{-4} t}{2.30}$; $t = 2110$ a

34. (a) $3\,{}^{4}_{2}\text{He}^{2+} \rightarrow {}^{12}_{6}\text{C}^{6+}$
    (b) $\Delta m = 11.996\,71 \text{ g} - 3(4.001\,50 \text{ g}) = -7.79 \times 10^{-3}$ g
        $\Delta E = -7.01 \times 10^8$ kJ (exothermic)

35. (a) ${}^{210}_{84}\text{Po} \rightarrow {}^{4}_{2}\text{He} + {}^{206}_{82}\text{Pb}$
    (b) $k = 0.693/140 \text{ d} = 4.95 \times 10^{-3} \text{ d}^{-1}$
    (c) $\log \dfrac{100}{10} = 1.0 = \dfrac{4.95 \times 10^{-3} t}{2.30}$; $t = 460$ d

*36. For a listing of isotopes and half-lives, consult a handbook of chemistry or a textbook of nuclear chemistry.

*37. First order reactions exhibit the property we are interested in; a given fractional change (~19%) occurs in a constant time interval (1000 a) (compare with a 50% change for $t_{1/2}$).

   $\log \dfrac{1.00}{1.00 - 0.19} = \dfrac{k(1000)}{2.30}$; $k = 2.1 \times 10^{-4}$ a^{-1}

   $\log (X_0/X) = \dfrac{2.1 \times 10^{-4} (1500)}{2.30}$, $X/X_0 = 0.73$ (In either language, 73% is unchanged.)

   Experiment: use ${}^{14}_{6}\text{C}$ to date a manuscript in the parent tongue (which would have the same number of terms fundamental to both modern vocabularies). You need only find such a manuscript!

SELECTED READINGS

A more extensive treatment of nuclear chemistry is given in:

Harvey, B. G., *Nuclear Chemistry*, Englewood Cliffs, N. J., Prentice-Hall, 1965.

Nuclear reactors, fission, and fusion, are discussed in almost every issue of The Bulletin of the Atomic Scientists. *See, especially for discussions of reactor safety, the issues of October and November, 1974, and September 1975. More:*

Hammond, R. P., Nuclear Power Risks, *American Scientist* (March-April 1974), pp. 155–160.

McIntyre, H. C., Natural-Uranium Heavy-Water Reactors, *Scientific American* (October 1975), pp. 17–27.

Weinberg, A. M., The Maturity and Future of Nuclear Energy, *American Scientist* (January-February 1976), pp. 16–21.

On the synthesis of the elements:

Selbin, J., The Origin Of the Chemical Elements, *Journal of Chemical Education* (May 1973), pp. 306–310.

On the application and misapplication of some nuclear reactions:

Hersey, J., *Hiroshima*, New York, Knopf, 1946.

Wahl, W. H., Neutron Activation Analysis, *Scientific American* (April 1967), pp. 68–82.

Pioneering the artificial isotope syntheses:

Impact: Interview with Glen T. Seaborg, *Journal of Chemical Education* (February 1975), pp. 70–75.

25
POLYMERS, NATURAL AND SYNTHETIC

QUESTIONS TO GUIDE YOUR STUDY

1. What is a polymer? Name a few polymeric substances.

2. How do polymeric substances differ from the molecular and ionic substances studied so far? What are the differences in atomic-molecular structure? In physical and chemical properties of bulk samples?

3. What kinds of intermolecular forces are present in polymers? How do they determine physical properties?

4. What kinds of substances may react to form polymers? Are there correlations you can make with the Periodic Table?

5. What energy effects accompany polymer formation? What are the extents, rates, and mechanisms of such reactions? To what degree have we learned to control these processes on an industrial scale?

6. How would you experimentally determine the composition and structure of a polymer (e.g., the sequence of bonded atoms and their three-dimensional arrangement in a protein)?

7. What conditions (T, P, concentration . . .) prevail during polymerization reactions in biological systems? How do enzymes catalyze these reactions?

8. Do proteins and other polymers spontaneously form from simpler molecular units? Is life itself a spontaneous process?

9. Can we begin to explain the properties of living organisms (such as growth and reproduction, mutation, disease, and death) in molecular terms?

25–Polymers, Natural and Synthetic

10. What kinds of answers can we now give as to how life may have begun? How evolution occurs?

11.

12.

YOU WILL NEED TO KNOW

Concepts

1. How to draw Lewis structures, predict molecular geometry, and the existence of *cis-trans* isomers — Chapter 8.
2. How to predict the kinds of intermolecular forces and their effects on physical properties — Chapter 9.
3. Some of the classes of organic compounds and their characteristic reactions — Chapter 10.

Math

1. How to work stoichiometric problems — Chapter 3.
2. How to relate bond energies and heats of reaction — Chapter 4.

CHAPTER SUMMARY

So far, most of our discussion of chemical principles has been illustrated by the properties of substances which are constituted of small molecules or ions. We now apply our general chemistry background to a broad class of substances where the unit of structure is a molecule perhaps a thousand times, or more, larger. The chemistry of natural and synthetic polymers brings together such topics as those listed above, and:
- methods of separation and identification — Chapter 1.
- molecular mass determination — Chapters 2, 12.
- methods and results of structure determination — Chapters 8 (this Guide), 11.
- principles of solubility — Chapter 12.
- enzyme catalysis — Chapter 16.

To illustrate the unification of chemical principles seen in this chapter, consider those natural polymers so important to living organisms. Proteins and carbohydrates, along with other organic compounds, constitute some 30% of your body mass. Despite the complexity of most of these substances and the reactions they take part in, we can note here several simplifications.

1. Though debate continues in the background, the nature of life seems to be explicable in terms of known principles of chemistry and physics. (In particular, no new chemical principles are needed in this chapter.) The molecular interpretation of biological structures and processes is already coming of age. For example: reaction pathways, the energy and material flows, have been discovered for most of the reactions involving the major constituents of living organisms.
2. Certain basic features are known to be common to all organisms: molecules with very similar architectures perform very similar functions, whether in one organism or another, plant, animal, or microbe. For example: the hemoglobin of man is built much like that of the horse and that of the gorilla, and all three serve to carry oxygen. There are really only a few kinds of biological molecules, and they are common to all life as we know it. (The usual classification lists fats and nucleic acids, as well as the proteins and carbohydrates, as the major components.)
3. Much of the chemistry of life occurs in dilute aqueous mixtures (and so rules out the need for studying many kinds of substances incompatible with water), at or near room temperature. The reactions generally, if not always, involve enzyme catalysis, and often oxidation-reduction as well.
4. Many, if not most, reactions are interlocked (coupled) with others. The energy released in one reaction is, at least in part, used to drive another reaction. For all coupled reactions, there is a net decrease in free energy. There is no creation of order in the living organism that doesn't occur simultaneously with a larger disordering of the surroundings.

In addition to being one of the most active research areas in recent years, molecular biology has been perhaps *the* area of cooperation among the various "compartments" of science. Working for perhaps a dozen man-years in elucidating the amino acid sequence for a single small protein, a team of physicists, chemists, and biologists have freely shared the tools and theories perfected individually, while training one another in the problems of their new frontier.

What are some of the unresolved or incompletely solved problems?
1. *Structures* are known in three-dimensional detail for only a handful of biological molecules. We know most other structures only in sketchy outline. All evidence so far indicates that all the biological

340 • 25–Polymers, Natural and Synthetic

properties of a molecule are determined by its primary structure (the number and kinds of atoms present and their bonding sequence).

2. *Molecular functions* and reaction mechanisms have been vaguely described for just a few representative compounds. Again, the details need to be discovered before a thorough understanding can be achieved.

3. *Comparative studies* of molecules serving similar functions need to be pursued in order to further describe the mechanism by which evolution occurs, by which only a few types of molecules seem to have survival value. The current work on compounds associated with fossils, as well as the abiological syntheses of compounds associated with living organisms, should continue to shed light on the possible origins of life itself.

4. Humanistic and moral questions need to be raised: most chemists consider as inevitable the deliberate *design* of organisms, as well as modification of organisms already alive. All of us need to know where we may be going.

BASIC SKILLS

This chapter is primarily descriptive; few new concepts are introduced.

1. **Given the formula of a monomer, write a structural formula showing part of:**
   a. **an addition polymer.**

The principles involved are demonstrated in Examples 25.1 and 25.2 and illustrated in Figure 25.1. See Problems 25.1(a), 25.8, 25.9, and 25.13, as well as the related answered problems.

   b. **a condensation polymer.**

This skill applies not only to polyesters and polyamides, but also to those polymeric substances classified as carbohydrates and proteins. Examples using this skill include 25.4 and 25.5. See also Figures 25.5 and 25.6. Problems include 25.2 and 25.12.

Conversely, sketch the monomer, given the structural formula for a portion of the polymer. Note that more than one monomer may be required, particularly in the case of a condensation polymer. See Problem 25.10.

2. **Given the formulas of one or more amino acids, write the structural formulas of all possible small polypeptides derived from the amino acid(s).**

Basic Skills • 341

This skill is illustrated in Example 25.7 and is required to work Problems 25.4 and 25.19.

3. **Calculate ΔH per mole of monomer for any polymerization, given bond energies (Table 4.2) and the formula(s) of the monomer(s).**

This skill is used in Example 25.3 and in Problems 25.1(b) and 25.17.
Additional stoichiometric calculations are required to work Problems 25.3, 25.20, and 25.21.

4. **Given the structural formula for a molecule, state whether or not it will show optical isomerism.**

At least for our purposes, we can expect to find optical isomers only if there is one or more asymmetric carbon atom in the molecule (i.e., a carbon atom which is bonded to four different groups). Thus, we expect the following molecules to show optical isomerism:

$$\underset{Br}{\overset{H}{Cl-\overset{|}{\underset{|}{C}}-OH}}, \quad \underset{Cl\ H}{\overset{H\ H}{H_3C-\overset{|}{\underset{|}{C}}-\overset{|}{\underset{|}{C}}-OH}}, \quad \underset{OH\ H}{\overset{H\ H}{H_3C-\overset{|}{\underset{|}{C}}-\overset{|}{\underset{|}{C}}-OH}}, \quad \underset{H\ H}{\overset{OH\ OH}{H_3C-\overset{|}{\underset{|}{C}}-\overset{|}{\underset{|}{C}}-CH_3}}$$

but not the following:

$$\underset{Cl}{\overset{H}{Cl-\overset{|}{\underset{|}{C}}-OH}}, \quad \underset{Cl}{\overset{H}{H_3C-\overset{|}{\underset{|}{C}}-H}}, \quad \underset{H}{\overset{H}{H_3C-\overset{|}{\underset{|}{C}}-CH_3}}, \quad \underset{H\ OH}{\overset{H\ OH}{H_3C-\overset{|}{\underset{|}{C}}-\overset{|}{\underset{|}{C}}-CH_3}}$$

See Figure 25.4 and the text discussion. The skill is required for Problems 25.14 and 25.15.

5. **Describe the preparation and/or properties of several polymers.**

See, for example, the text discussion of polyethylene, polyvinyl chloride, and rubber and of nylon and silicone oil. Again, see the discussion of the properties of carbohydrates and proteins, as they depend on functional groups, and on hydrogen bonding. Also, be able to describe the functioning of an enzyme in its role as a catalyst. (A review of parts of Chapter 16 would be helpful here.)
Problems calling for such discussion include 25.5–25.7, 25.16, 25.23, and the related problems with answers.

SELF-TEST

True or False

1. Most synthetic polymers are solids at 25°C, 1 atm. ( )

2. Formation of high MM polymers is limited to organic compounds. ( )

3. Most organic polymers retain their useful properties over a narrow range of temperatures. ( )

4. Carbohydrate polymers are synthesized only by plants. ( )

5. The net photosynthetic reaction, $CO_2(g) + H_2O(g) \rightarrow$ glucose(s) $+ O_2(g)$, involves reduction of carbon. ( )

6. The α-amino acids might generally be expected to form chelates with metal atoms. ( )

7. The water solution of any amino acid which contains only one basic functional group and one acidic group is expected to have a pH of 7. ( )

8. The hydrolysis of a polyester is the reverse of a condensation reaction. ( )

9. The formation of a polypeptide from individual amino acid molecules is a condensation polymerization. ( )

10. $\Delta G > 0$ for the synthesis of an enzyme from amino acids. The fact that the synthesis of enzymes occurs in a living organism is a contradiction of the laws of thermodynamics. ( )

Multiple Choice

11. The method of choice for determining the molecular mass of a polymer is ( )
    (a) osmotic pressure
    (b) gas density
    (c) freezing point lowering
    (d) direct weighing of a single molecule

12. The best solvent for a polypeptide is likely to be ( )
    (a) CCl_4
    (b) CH_3-O-CH_3
    (c) H_2O
    (d) HF

13. To separate a mixture of monosaccharides, you would ( )
probably use a(n)
    (a) centrifuge
    (b) chromatograph
    (c) mass spectrometer
    (d) electrolytic cell

14. The linkage between monomers in a polyester is ( )
    (a) C—O—C
    (b) $\underset{\underset{O}{\|}}{C}$—O—C
    (c) $\underset{\underset{O}{\|}}{C}$—$\underset{\underset{H}{|}}{N}$—C
    (d) a hydrogen bond

15. Addition polymers are expected to form from ( )
    (a) F_3C-CF_3
    (b) $F_2C=CF_2$
    (c) CF_4
    (d) F_2

16. Condensation polymers are expected to form from ( )
    (a) CH_3NH_2
    (b) $HCOOH$
    (c) $H_2N-CH_2(CH_2)_4CO_2H$
    (d) $HO-CH_2CH_2-OH$

17. The ability to form polymers depends on the presence of ( )
    (a) carbon atoms
    (b) unpaired electrons
    (c) double bonds
    (d) two at least potentially reactive sites on a molecule

18. Most synthetic polymers are ultimately derived from ( )
    (a) wood
    (b) natural gas
    (c) silicates
    (d) petroleum

19. Polymeric materials which possess highly regular molecular ( )
packings tend to
    (a) be crystalline
    (b) have high density
    (c) have high melting point
    (d) all of these

20. Many polymeric carbohydrates are condensation polymers ( )
of
    (a) glucose
    (b) amino acids
    (c) ethylene
    (d) phenol

21. The reaction of HO—R—OH and $HO_2C-R'-CO_2H$ is ( )
expected to give a(n)
    (a) polyamide
    (b) polyester
    (c) polypropylene
    (d) polysaccharide

22. Suppose the metabolism of glucose could be represented ( )
simply as

$$C_6H_{12}O_6(s) + 6\ O_2(g) \rightarrow 6\ CO_2(g) + 6\ H_2O(l).$$

The *maximum* amount of work that this reaction could do is equal in magnitude to:
    (a) ΔH (b) ΔG
    (c) $\dfrac{\Delta H - \Delta G}{T}$ (d) $T\Delta S$

23. Which of the following carbohydrates cannot be directly ( )
utilized by the human body as a source of energy?
    (a) glucose   (b) sucrose
    (c) glycogen   (d) cellulose

24. Which one of the following formulas would you choose to ( )
represent an average composition of protein?
    (a) CH_2O (b) $C_{57}H_{110}O_6$
    (c) CH_7NO (d) $C_9H_{13}SN_2O_2$

25. The geometry associated with the atoms of the peptide ( )
link,

$$\text{C} - \overset{\overset{\displaystyle O}{\|}}{\text{C}} - \overset{\overset{\displaystyle H}{|}}{\text{N}} - \text{C},$$

is observed to be
    (a) linear
    (b) tetrahedral
    (c) planar
    (d) variable, depending on the rest of the chain

26. In a 0.10 M solution of glycine, $H_2C(NH_2)CO_2H$, at a pH ( )
of 10, the most abundant species (next to water) is
    (a) OH^- (b) $H_2C(NH_3)CO_2$
    (c) $H_2C(NH_3)CO_2H^+$ (d) $H_2C(NH_2)CO_2^-$

27. The maximum number of different tripeptides that can be ( )
formed from three different amino acids, using one residue of each, is
    (a) 2   (b) 3
    (c) 4   (d) 6

28. The maximum number of different tripeptides that can be ( )
made from three different amino acids, using any number of residues of each, is
    (a) 2   (b) 3
    (c) 6   (d) 27

29. As the temperature is increased, the rate of an enzyme- ( )
catalyzed reaction first increases and then decreases. The decrease can be explained in the following way:
    (a) All reactions achieve a maximum rate at some temperature.

(b) The molecules acted on by the catalyst decompose spontaneously at the high temperature.
(c) Intramolecular forces (e.g., H bonding) in the enzyme molecule begin to break down.
(d) All catalyzed reactions behave in this unaccountable manner.

30. The rate of an enzyme-catalyzed reaction is generally found ( ) to increase with the concentration of the reactant substrate up to some maximum and then level off. This levelling-off is probably due to
   (a) decomposition of enzyme
   (b) a change in enzyme conformation
   (c) molecules of reactant get in the way of each other at high concentrations
   (d) all the "active sites" on the enzyme molecules are occupied

31. Which one of the following species would you expect *not* to ( ) show optical activity?
   (a) $CH_2(NH_2)CO_2H$
   (b) $CH_3CH(NH_2)CO_2H$
   (c) $CH_3\overset{OH}{\underset{|}{C}}H-\overset{O}{\underset{||}{C}}H$
   (d) $CH_3CH_2\overset{OH}{\underset{|}{C}}HCH_3$

32. The "backbone" in a protein molecule can assume only a ( ) limited number of conformations. This is a result of
   (a) hydrogen bonding
   (b) relative solubilities of attached groups
   (c) the geometry of the peptide link
   (d) all of the above

33. The argument for evolution gains support from which of the ( ) following?
   (a) Proteins and carbohydrates are not sufficiently stable to be found in fossils.
   (b) The function of a particular enzyme in man is served by enzymes of similar structure in other organisms.
   (c) The "active site" in an enzyme is the locale for most mutations.
   (d) Proteins are found in all organisms.

Problems

34. Identify the monomers from which the following polymers were formed.

346 • 25-Polymers, Natural and Synthetic

(a)
$$-\underset{\underset{Cl}{|}}{\overset{\overset{H}{|}}{C}}-\underset{\underset{H}{|}}{\overset{\overset{Cl}{|}}{C}}-\underset{\underset{Cl}{|}}{\overset{\overset{H}{|}}{C}}-\underset{\underset{H}{|}}{\overset{\overset{Cl}{|}}{C}}-\underset{\underset{Cl}{|}}{\overset{\overset{H}{|}}{C}}-$$

(b) $-\underset{\underset{O}{\|}}{C}-(CH_2)_2-\underset{\underset{O}{\|}}{C}-O-(CH_2)_2-O-\underset{\underset{O}{\|}}{C}-(CH_2)_2-\underset{\underset{O}{\|}}{C}-$

(c) $-\underset{\underset{}{\overset{H}{|}}}{N}-(CH_2)_2-\underset{\underset{H}{|}}{\overset{\overset{O}{\|}}{C}}-N-(CH_2)_2-\underset{\underset{O}{\|}}{\overset{\overset{H}{|}}{C}}-N-$

35. Vinyl alcohol, $CH_2=CHOH$, forms a polymer that is rather soluble in water.
    (a) Sketch a portion of the polymer.
    (b) Estimate ΔH for the polymerization of a mole of alcohol.
    (c) Explain the solubility, unusual among polymers.
    (d) Estimate the number of monomer residues in a solution containing $1.0\ g/dm^3$ that gives an osmotic pressure of 1.0 kPa at 25°C.

*36. Polymer chemistry is a very active, densely populated, area of research. Interview a polymer chemist to find out why.

SELF-TEST ANSWERS

1. **T** (Reflecting strong intermolecular attractions – Chapter 9.)
2. **F** (Though in the minority, inorganic polymers are known: glass fiber, silicones, plastic sulfur, etc.; see Readings.)
3. **T**
4. **F** (Glycogen is but one exception.)
5. **T** (What are the oxidation numbers? H_2O is oxidized to O_2 – see Chapter 22.)
6. **T** (Like $\begin{matrix}C-O\\|\\C-N\end{matrix}\,$M. What is required of such a ligand? See Chapter 21.)
7. **F** (Their strengths may not be the same.)
8. **T** (H_2O is inserted at each ester linkage, giving alcohol and acid functional groups.)
9. **T** (H_2O is eliminated for each peptide link.)
10. **F** (First, the biosynthesis doesn't begin with free amino acids but with complex molecules such as proteins. Second, other

Self-Test Answers • 347

reactions, for which $\Delta G < 0$, occur that supply the work necessary for the synthesis.)

11. a (Most sensitive — see Chapter 12.)
12. c (A nondestructive solvent for these molecules that contains lots of polar groups. Principles of solubility were discussed in Chapter 12.)
13. b (Chapter 1.)
14. b (Ester formation was discussed in Chapter 10.)
15. b (Teflon is the commercial product.)
16. c (A nylon would result; draw part of the structure and compare to Nylon 66.)
17. d (Either b or c may be involved, or two reactive functional groups.)
18. d
19. d (Solid state structure and physical properties are related in Chapters 9 and 11.)
20. a
21. b
22. b (Chapter 14; but of course, our metabolic machine isn't 100% efficient.)
23. d
24. d
25. c (Consider the resonance structure $\begin{smallmatrix}C\\ \diagdown\\ O\end{smallmatrix}C=N\begin{smallmatrix}H\\ \diagup\\ \diagdown C\end{smallmatrix}$. See Chapter 8 on predicting geometry.)
26. d (What is [OH$^-$]? See Chapter 19.)
27. d (Each amino acid residue could be at the —NH$_2$ end or at the —CO$_2$H end.)
28. d
29. c (Its ability to function strongly depends on its shape.)
30. d (Recall the discussion of zero order, surface reaction mechanism, Chapter 16.)
31. a (No asymmetric carbon.)
32. d
33. b (The number of differences in structure of such related enzymes is taken to be a measure of their time separation on the evolutionary tree.)
34. (a) ClHC=CHCl
    (b) HO—(CH$_2$)$_2$—OH and HO$_2$C—(CH$_2$)$_2$—CO$_2$H
    (c) H$_2$N—(CH$_2$)$_2$—CO$_2$H
35. (a)
$$-\overset{\overset{\displaystyle H}{|}}{\underset{\underset{\displaystyle H}{|}}{C}}-\overset{\overset{\displaystyle H}{|}}{\underset{\underset{\displaystyle OH}{|}}{C}}-\overset{\overset{\displaystyle H}{|}}{\underset{\underset{\displaystyle H}{|}}{C}}-\overset{\overset{\displaystyle H}{|}}{\underset{\underset{\displaystyle OH}{|}}{C}}-$$

(b) Per monomer, C=C is broken and C—C is formed at either end:

$$\Delta H = BE(C=C) - 2\,BE(C-C) = 598 - 2(347) = -96\text{ kJ}$$

(c) Hydrogen bonding

(d) $\pi = \dfrac{g/GMM}{V} RT$

$$GMM = \frac{gRT}{\pi V} = \frac{1.0\,(8.31)(298)}{(1.0)(1.0)} = 2500$$

2500 = MM of $(CH_2CHOH)_n$; $n(44/\text{monomer}) = 2500$, $n = 57$

SELECTED READINGS

Polymers, natural and synthetic, are considered in:

Allcock, H. R., Inorganic Polymers, *Scientific American* (March 1974), pp. 66–74.
Materials, *Scientific American* (September 1967), the entire issue.
Morton, M., Polymers — Ten Years Later, *Chemistry* (October 1974), pp. 11–14.
Uhlmann, D. R., The Microstructure of Polymeric Materials, *Scientific American* (December 1975), pp. 96–106.

Biological molecules, including DNA, and chemical evolution, are discussed in almost any issue of Scientific American, *and:*

Calvin, M., *Chemical Evolution*, New York, Oxford, 1969.
Calvin, M., Chemical Evolution, *American Scientist* (March-April 1975), pp. 169–177.
Watson, J. D., *Molecular Biology of the Gene*, Menlo Park, Ca., W. A. Benjamin, 1970.